HUMAN FACTORS FOR PILOTS

Human Factors for Pilots

Second Edition

ROGER G GREEN
HELEN MUIR
MELANIE JAMES
DAVID GRADWELL
ROGER L GREEN

Ashgate

Aldershot • Burlington USA • Singapore • Sydney

Published by
Ashgate Publishing Limited
Gower House
Croft Road
Aldershot
Hants GU11 3HR
England

Ashgate Publishing Company
131 Main Street
Burlington, VT 05401-5600 USA

Ashgate website:http://www.ashgate.com

Reprinted 1997, 1998, 1999, 2001

A CIP catalogue record for this book is available from the British Library

ISBN 0 291 39827 8

Typesetting : Melanie James
Illustrations : Roger G Green and David Campbell

Printed in Great Britain by MPG Books Ltd, Bodmin, Cornwall

CONTENTS

PART II

BASIC AVIATION PSYCHOLOGY

PART III

STRESS, FATIGUE, AND THEIR MANAGEMENT

PART IV

THE SOCIAL PSYCHOLOGY AND ERGONOMICS OF THE FLIGHT DECK

Sections marked ✈ are not included in the UK Private Pilot Syllabus

PREFACE

'Human Factors' is a strange and possibly ungrammatical name for a discipline or study. Nevertheless, it has come to be used to encompass all of those considerations that affect man at work. No form of work has been studied more closely in this regard than that of flying, yet the existing knowledge has not generally been available to pilots in a form designed especially for them.

This handbook sets out to remedy this situation. It does not purport to deal with the subject exhaustively and in depth, though there are a number of books in the reading list at the back of this book that do. Instead, it has two functions. The first is to provide a reasonably comprehensive but concise outline of the available human factors knowledge about flying in the hope that it will prove useful and interesting to any practising pilot. The second function is to provide this knowledge in a way that follows very closely the syllabus of the UK Civil Aviation Authority's (CAA) Human Performance and Limitations examinations for both professional and private pilots. Although the private pilot's syllabus requires a narrower range of subjects to be studied, and in less detail, than the professional syllabus, this handbook covers both requirements, with syllabus variations being indicated in the contents pages. Any student who has followed and absorbed the material that follows, should have no difficulty with the examination.

The four major sections into which this handbook is divided contain material from psychology, physiology, and medicine. All of these subjects are well known for the use of unnecessarily arcane and esoteric language, a problem that the authors of this handbook have tried hard to avoid. The authors are all professionally employed to research or to practise in their areas of expertise.

Roger G Green and Melanie James are employed at the DERA Centre for Human Sciences, David Gradwell at the RAF School of Aviation Medicine, Helen Muir at Cranfield University, and Roger L Green is an aviation medicine consultant to both the CAA and British Airways. Special thanks are also owed to Rowland Peacock and Alan Laing of Cranfield for their contributions.

A word of apology must be made to all female readers. Throughout this handbook the pilot has been referred to as 'he'. This is simply to avoid the cumbersome repetitive use of 'he or she' in the absence of a suitably neutral pronoun. It is true that the overwhelming majority of present commercial pilots are male, but this is a state of affairs that is certainly changing.

In the five years since this book was first published, many courses and publications addressing Human Factors issues in aviation have appeared, and this growth of interest in the area must be welcomed. Nevertheless, the basics have not changed radically in this time and it is hoped that the readers of this book will find it as useful in this second edition version as when it was originally published.

Roger G Green

PART I

BASIC AVIATION

PHYSIOLOGY

AND

HEALTH

MAINTENANCE

1a BASIC PHYSIOLOGY AND THE EFFECTS OF FLIGHT

Introduction

This chapter will outline the relationships between human physiology and flight with emphasis on the limitations imposed by the body on pilot performance. Some areas of basic physics will also be revised insomuch as they apply to flight.

1a.1 Composition of the Atmosphere

The atmosphere is composed of a number of concentric 'shells' around the Earth but for conventional aviation purposes it is necessary to understand only the physical relationships that exist in the innermost layer of the atmosphere, the troposphere. This region, which extends to an altitude of 30 000ft at the poles and 60 000ft at the equator, is characterized by a relatively constant decline in temperature with increasing altitude at a rate of 1.98°C/1000ft. The air itself has weight, therefore compresses towards the earth, and as a result density (and therefore atmospheric pressure) increases with decreasing altitude in an exponential manner. Small increases in height at low altitude therefore cause a much greater change in pressure than the same change in height at high altitude (see Figure 1a.1).

The air through which an aircraft flies is composed of a mixture of gases of remarkably constant proportions, ie:

Oxygen 21%

Nitrogen 78%

Other gases 1%

The other gases include argon, water vapour, and carbon dioxide. For practical purposes however, atmospheric air can be considered to be a mixture of two dry gases, oxygen (21 per cent) and nitrogen (79 per cent). Similarly, there is a need to standardize the description of the physical relationships in the atmosphere and this is achieved by the adoption of a 'standard atmosphere'. The most widely accepted is the ICAO standard atmosphere which defines the composition of the air, its sea-level density and pressure, acceleration due to gravity, and the temperature and pressure profiles.

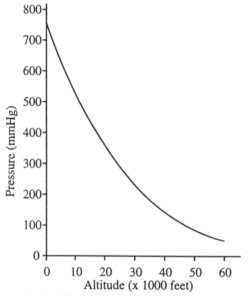

Figure 1a.1. Altitude/pressure relationship of the atmosphere

Gas Laws

The gas laws of particular relevance in aviation medicine are Boyle's law which relates pressure to volume, Charles' law which relates temperature to volume and Dalton's law of partial pressures. Dalton's law expresses the relationship between the partial pressure of a component gas of a mixture to its fractional concentration. From this, the partial pressure of oxygen in the atmosphere can be derived for any altitude since the pressure at that altitude can be measured and the concentration of oxygen in atmospheric air is constant. As will be seen this is of great significance to man in his ascent through the atmosphere.

1a.2 The Human Requirement for Oxygen

To live, man must produce energy from the food he eats. He does this by chemical oxidation of simple food products in the tissues. This respiratory process requires the delivery of oxygen to every living cell and the carriage away from it of the waste product carbon dioxide. Thus respiration can be considered to be the exchange of respiratory gases, oxygen and carbon dioxide, between the environment and the tissues. Transport within the body is carried out by the blood, and because the body has only a very small store of oxygen there is a constant demand for replenishment to take place. In contrast, the body has quite a large store of carbon dioxide and uses this as a means of moderating the balance of acids and bases within the blood and tissues.

The Lungs and the Transport of Oxygen

Air is drawn into the lungs by outward movement of the chest wall and downward movement of the diaphragm resulting in a fall of pressure inside the chest. It is expelled from the lung by the generally passive process of muscular relaxation allowing the chest wall to fall and the diaphragm to relax, pushed back towards the chest by the abdominal structures compressed during inhalation. Air enters through the nose and mouth and passes down the trachea to the bronchial tree. These ever dividing passageways terminate at alveoli, very fine sac-like structures where blood in the alveolar capillaries is brought into very close proximity with oxygen molecules (see Figure 1a.2). Under the

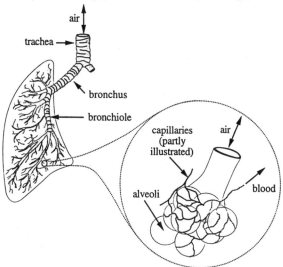

Figure 1a.2. Passage of air in the lungs

influence of a pressure gradient, oxygen diffuses across the capillary membrane from the alveolar sac into the blood. There it is taken up by the protein molecule haemoglobin for transport around the body. Haemoglobin, found within the red blood cell, is a complex and highly specialized oxygen transport system that allows far more oxygen to be carried by blood than could be achieved by simple solution. Haemoglobin also has the property of remaining bound to oxygen molecules until it enters an area of very low oxygen tension where the oxygen is released to diffuse into the tissues. Carbon dioxide is also carried in the blood but largely dissolved as carbonic acid and bicarbonate.

Breathing provides an exchange of respiratory gases between the environment and the blood; the rate and depth of breathing are adjusted to meet the enormous changes in the consumption of oxygen and the elimination of carbon dioxide. Chemical receptors in the brain monitor the levels of both gases and make necessary changes in the respiratory pattern. A rise in carbon dioxide concentration or a fall in oxygen causes an increase in respiratory ventilation. However, the healthy body is more sensitive to changes in carbon dioxide than to oxygen.

Circulation

A block diagram of the human circulation is shown in Figure 1a.3. Red blood cells loaded with oxygen rich haemoglobin pass from alveolar capillaries into pulmonary veins and then to the left atrium of the heart. From there blood passes to the left ventricle from which it is forcibly expelled around the body via the aorta and the arterial branches. The blood travels through arterial branches of ever increasing number but reducing diameter until it arrives in the capillaries, the finest, thin walled blood vessels. There oxygen is in close proximity to the tissues and unlatches from haemoglobin. The oxygen molecules diffuse down a pressure dependent gradient across the cell walls into the respiring tissues. Carbon dioxide is picked up in exchange, and the capillary blood passes on into the veins. The weight of blood in the upper body and muscle pump action in the limbs drive blood in the low pressure venous system back towards the heart. Valves in the veins prevent blood travelling in the wrong direction, until ultimately venous blood arrives back at the right side of the heart. From the right ventricle it is pumped to the lung and the whole process starts again.

1a.3 Effects of Reduced Ambient Pressure

The passage of oxygen from the alveoli into red blood cells is dependent on a pressure gradient and so it is to be expected that if the driving pressure falls the movement into the blood is impaired. With ascent, the partial pressure of oxygen falls in parallel with atmospheric pressure. Although the strong affinity of haemoglobin for oxygen provides some degree of protection this fails above an altitude of 10 000ft at which the partial pressure of oxygen in the alveoli is 55mmHg, compared with 103mmHg at sea-level. Even as low as 8 000ft it is possible to demonstrate some impairment of higher mental processes (particularly learning ability and the performance of novel tasks), but for

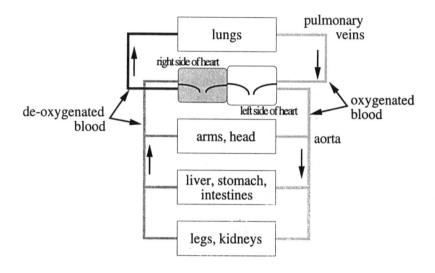

Figure 1a.3. Block diagram of the human circulation

healthy individuals ascent to 10 000ft should not produce noticeable symptoms. Above 10 000ft it is necessary to increase the concentration of oxygen in the inspired gas to maintain a partial pressure of oxygen in the lung sufficient to avoid hypoxia (inadequate oxygen).

Oxygen Requirements and Cabin Pressurization

The addition of an increasing proportion of oxygen into a pilot's breathing gas (airmix) can be used to avoid hypoxia until the point is reached at which he is breathing 100 per cent oxygen to maintain an alveolar partial pressure equivalent to that found breathing air at sea-level (103mmHg). This point is reached at an altitude of 33 700ft. However, as is noted above, it is permissible to allow the alveolar partial pressure of oxygen to fall to 55mmHg. With 100 per cent oxygen as the breathing gas this permits an altitude of 40 000ft to be achieved. To go any higher than this it is necessary to deliver the oxygen under pressure, ie Positive Pressure Breathing (PPB). The PPB technique has been used considerably in the military field as a 'get you down' facility and is also available to Concorde flight deck crews, but is very tiring and requires practice to perfect.

Thus in summary the requirements for oxygen are as follows:

0–10 000ft	air only is required, some higher function impairment.
10 000–33 700ft	increasing percentage of oxygen in breathing gas.
33 700–40 000ft	100 per cent oxygen
Above 40 000ft	100 per cent oxygen under pressure.

All of these heights relate to the actual altitude to which the pilot is exposed but cabin pressurization systems ensure that the effective altitude to which the aircrafts' occupants are actually exposed is much lower than that at which they are flying. The physiological ideal would be to produce the equivalent of sea-level but this is impractical because of weight and aircraft strength considerations. Therefore compromises are made in terms of cabin pressure differentials. The pressurization of a commercial airliner at 30 000ft produces an environment inside the hull in which the pressure altitude is approximately 6 000ft. The cabin conditioning system also maintains an acceptable 'shirt sleeve' temperature inside the cabin. The pressure differential across the cabin wall may be in the region of 9psi at its greatest, and if the aircraft had to fly higher than expected the cabin altitude would, of necessity, rise to prevent this differential being exceeded.

1a.4 Hypoxia and Hyperventilation

Hypoxia is the term used to describe the condition which occurs when the oxygen available to the tissues is insufficient to meet their needs. There are many causes, but in aviation the greatest risk of hypoxia arises as a result of ascent to altitude with its associated fall in ambient pressure.

The earliest symptoms of hypoxia are related to higher mental function because brain tissues are the most sensitive to lack of oxygen. The effects seen are similar to the effects of alcohol but the rate of onset is greatly influenced by the altitude to which an individual is exposed. Thus at 20 000ft the signs and symptoms are relatively slow in onset, whereas at 30 000ft they may be so quick that subtleties are not seen. Furthermore, although the symptoms are similar in all individuals, there are marked variations with everyone having their own symptom pattern that tends to recur on each exposure to hypoxia. The classical signs and symptoms may be summarized as follows:

Apparent personality change – change in outlook and behaviour with euphoria or aggression and loss of inhibitions.

Impaired judgement - loss of self-criticism with the individual unaware of reduced performance.

Muscular impairment - finely coordinated movements become difficult through slow decision making and poor fine muscular control.

Memory impairment - short term memory is lost early, making drills difficult to complete unless trained into long-term memory.

Sensory loss - vision, especially for colour, is affected early, then touch, orientation, and hearing are impaired.

Impairment of consciousness - as hypoxia progresses the individual's level of consciousness drops until he becomes confused, then semi-conscious, and unconscious. Unless he is rescued he will die and at high altitude death can occur within a few minutes.

An individual who has become hypoxic at altitude is likely to be cyanosed, that is his lips and fingertips develop a blue tinge because so much haemoglobin in the circulation is in the deoxygenated state. He will hyperventilate (overbreathe) in an effort to get more oxygen, but this is of relatively little benefit when in an environment of low ambient pressure.

Susceptibility to hypoxia is increased by many factors, the most important of which are:

Altitude - the greater the altitude the greater the hypoxia and the more rapid its progression.

Time - the longer the time of exposure to altitude, the greater the effect.

Exercise - exercise increases the demand for oxygen and hence increases the degree of hypoxia.

Cold - cold makes it necessary to generate more energy which in itself increases the demand for oxygen and hence increases the degree of hypoxia.

Illness - similarly illness increases the energy demands of the body.

Fatigue - fatigue lowers the threshold for hypoxia symptoms.

Drugs/alcohol - alcohol has very similar effects on the body as hypoxia and therefore reduces tolerance of altitude. Many other drugs have adverse effects on brain function and therefore further impair function under hypoxic conditions.

Smoking - carbon monoxide, produced by smoking, binds to haemoglobin with a far greater affinity than oxygen and therefore has the effect of

reducing the available haemoglobin for oxygen transport, exacerbating any
degree of hypoxia.

Some of the these factors are unavoidable risks of flight but others can be
reduced by good personal habits and forethought. The importance of aircrew
being able to recognize hypoxia cannot be overstated. Armed with the
knowledge of the signs and symptoms of hypoxia, early identification of the
problem will allow the correct drills to be carried out before anyone is placed in
jeopardy, but it can be seen from the details given above that it is important that
those drills are well learnt and easily accomplished. Failure to do so can and has
lead to the loss of aircraft and the lives of all on board. In 1979 the crew of a
King Air 200 carried out a practice decompression over Exeter at an altitude
above 30 000ft and probably failed to recognize the rapid onset of hypoxia
before losing consciousness. The aircraft crashed in France several hours later
when it ran out of fuel, having flown in a corkscrew pattern for hundreds of
miles with the crew dead at the controls. Pilots must familiarize themselves with
the appropriate oxygen drills for the aircraft in which they are flying before
venturing to an altitude at which hypoxia can occur ie above 10 000ft.

The time available to a pilot to recognize the development of hypoxia and to do
something about it is termed the *time of useful consciousness*. This is not the
time to unconsciousness but is the rather shorter time from a reduction in
adequate oxygen tension until a specified degree of impairment, generally taken
as when the individual can no longer take the necessary steps to help himself.
The time interval involved may be 30 minutes at 18 000ft, 2-3 minutes at
25 000ft, 45-75 seconds at 30 000ft and 12 seconds at 45 000ft. The precise
time may be influenced by the factors listed above.

Hyperventilation, as has been noted above, is overbreathing. That is, breathing
in excess of the ventilation required to remove (or, more accurately, to maintain
a normal alveolar partial pressure of) carbon dioxide. Overbreathing reduces the
alveolar carbon dioxide, and thus induces changes in the acid-base balance of
the body that result in widespread symptoms. Although hypoxia causes
hyperventilation it is far from the only cause. Anxiety, motion sickness,
vibration, heat, high G, pressure breathing, and more, can all cause the
individual to suffer the symptoms of hyperventilation. Those symptoms are:

Dizziness

Tingling - especially in the hands, feet and around the lips.

Visual disturbances - particularly tunnelling or clouding.

Hot or cold feelings - which may alternate in time or site on the body.

Anxiety - establishing a vicious circle of cause and effect.

Impaired performance - pilot performance can be dramatically reduced
by hyperventilation.

Loss of consciousness - hyperventilation can lead to collapse but thereafter respiration returns to normal and the individual recovers, (unless, of course, the pilot crashes into the ground before recovery).

In the event of symptoms in flight it can be very difficult to distinguish between the effects of hypoxia and the effects of hyperventilation. The appropriate response of the pilot must be to assume the worst and, if he is at an altitude at which hypoxia is a possibility he must take that to be the cause. Thus he should carry out his hypoxia drills as discussed above. If symptoms occur at an altitude at which hypoxia is not a consideration, ie below 10 000ft he should regulate the rate and depth of respiration to reduce the overbreathing, and thereby restore to normal the acid-base balance in his blood and alleviate the symptoms.

Cabin Decompression

Loss of cabin pressurization can occur in flight. The rate of loss may be very slow with the crew recognizing the problem and making appropriate height reductions before the passengers are aware of anything amiss. Very occasionally there is a rapid decompression such as may result from the loss of a window or door, or terrorist action (in June 1990 a British Airways BAC 1-11 en route from Birmingham suffered the loss of the captain's windshield and a pilot was almost lost when sucked out of the cockpit). In that event the occupants are very abruptly exposed to the full rigours of high altitude with all the attendant risks of hypoxia, cold and decompression sickness (1a.6). The aircraft must rapidly descend to a safer altitude and all on board may need to use emergency oxygen to avoid hypoxia.

In cases of rapid decompression the altitude in the cabin may actually rise above that of the aircraft. This is brought about by an effect known as aerodynamic suction whereby air on the outside, passing quickly over the defect in the aircraft hull has a Venturi effect on the remaining air in the cabin. Therefore further air may be sucked out and the inside altitude rise still higher. The precise degree to which this may occur depends on the position of the defect in the airstream and the height of the aircraft, but may amount to as much as 5 000ft in pressure terms.

1a.5 Entrapped Gas and Barotrauma

One of the consequences of ascent to altitude is the application of Boyle's Law to gas contained within the body cavities. The Law states that pressure is inversely related to volume, providing the temperature remains constant. Body temperature is essentially constant and so on ascent, as pressure drops, the volume of gas within body cavities increases.

Middle Ear

Figure 1a.4 shows the anatomy of the ear. The middle ear is an air-filled cavity in communication with the nose and throat via the Eustachian tube. The walls of the Eustachian tube are soft and the nasal end acts as a flap valve. This allows expanding gas in the middle ear cavity to vent on ascent but on descent, with an increase in ambient pressure, this flap valve can stop air returning to the middle ear to equalize the pressure. This failure to restore the correct pressure inside the middle ear results in distortion of the ear drum giving rise to pain and injury known as otic barotrauma.

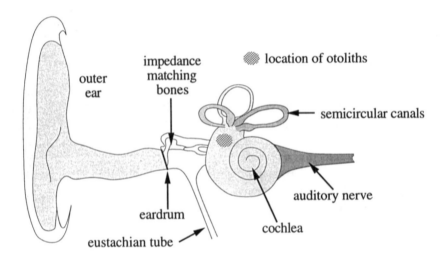

Figure 1a.4. The anatomy of the ear

Sinuses

The sinuses are cavities within the skull that help to make the voice resonant and in addition make the skull lighter than it would otherwise be. Sinuses are situated above the eyes, in the cheeks and at the back of the nose. In a very similar manner to the ears the sinuses can vent gas more easily on ascent than can re-enter on descent, and sinus barotrauma may occur as a result. This is especially true whilst the individual is suffering the symptoms of a cold or flu. Then, the soft tissues around the sinuses (and the Eustachian tube) expand and make the obstruction to inward passage of air on descent even more difficult. For this reason aircrew with a cold or flu must avoid flying whilst symptoms persist. Failure to do so can result in barotrauma of ears or sinuses that would make them unfit to fly for perhaps weeks or even longer.

Gastrointestinal Tract

The gastrointestinal tract is, in effect, a tube from mouth to anus, and gas collects at various points along that tube. When we eat, air is swallowed along with the food and as a result a bubble of gas collects in the stomach. On ascent this bubble will expand but can generally escape up through the mouth without difficulty. At the other end of the tube bacteria in the large bowel produce gas and this too will expand on ascent but can be vented through the anus. There remains the problem of gas occurring elsewhere in the tract, often the result of fermentation of particular foodstuffs, particularly beer, beans, and some highly spiced foods, such as curries. Gas may form in the small bowel that is too far from either entrance or exit to escape. Expansion of this gas causes stretching of the wall of the bowel which can be very painful, occasionally leading to the individual fainting. Aircrew likely to be exposed to great changes of pressure should learn from experience the foodstuffs to avoid before flight.

Teeth

Healthy teeth do not contain gas but gas pockets in and around teeth can occur either as a result of poor filling or from dental abscess formation at the apex of the tooth. If this occurs, ascent to altitude and the associated expansion of gas can cause tooth pain, termed 'aerodontalgia'. Good dental care and hygiene can prevent this problem quite easily.

Lungs

The lungs contain a very large volume of gas but are generally in easy communication with the outside. Therefore ambient pressure changes can readily be accommodated and no pressure differential will develop. The only potential risk arises from very rapid decompressions but, provided the individual breathes out during the decompression, lung damage is extremely rare.

1a.6 Decompression Sickness

Decompression sickness occurs in association with exposure to reduced atmospheric pressure and is characterized by the evolution of bubbles of nitrogen coming out of solution in body tissues. Although nitrogen is only poorly soluble there is enough present to give rise to problems under very particular circumstances. Ascents to altitude above 18 000ft and especially over 25 000ft are associated with a small incidence of decompression sickness unless certain steps are taken to avoid it.

The body is normally saturated with nitrogen. When ambient pressure is abruptly reduced on ascent some of this nitrogen comes out of solution as bubbles. The symptoms vary according to the site involved. Thus bubbles in the joints cause an aching pain termed the 'bends'. In aviation, the most commonly affected joints are the shoulder, the elbow, the wrist, the knee, and the ankle. Movement or rubbing the joint aggravates the pain but descent is usually associated with resolution of symptoms. Nitrogen bubbles in the skin cause the 'creeps', in the respiratory system the 'chokes', and in the brain the 'staggers'. Ultimately, the individual may collapse and in rare cases decompression sickness may occur, or persist, after descent and go on to cause death.

Decompression sickness (DCS) is made more likely the higher and the longer an exposure to altitudes above 18 000ft continues. Hypoxia and cold also increase the risk, as does being over-weight and an individual's own susceptibility. To a great extent the risk of DCS can be avoided by pre-oxygenation in which the pilot breathes 100 per cent oxygen for a period prior to high altitude exposure, reducing the body store of nitrogen as much as possible. If DCS does occur in flight immediate descent must be instituted and the aircraft should land as soon as possible. The patient should be put on 100 per cent oxygen and kept warm and rested. On landing he may need recompression to depth and other supportive medical treatment.

Flying and Diving

Decompression sickness is rare, but the incidence is greatly increased for individuals who have been diving shortly before a flight. In SCUBA diving air under pressure is used as a source of breathing gas and this increases the body's store of nitrogen. On subsequent ascent this may come out of solution, giving rise to decompression sickness. Therefore strict guidelines are set down concerning diving and flying. **Do not fly within 12 hours of swimming using compressed air and avoid flying for 24 hours if a depth of 30ft has been exceeded.** Similar rules apply following ground recompression experience. Failure to adhere to these rules results in incidents each year in which individuals develop decompression sickness in flight at cabin altitudes as low as 6 000ft.

1a.7 Effects of Acceleration

The ability of the body to cope with the effects of acceleration depends on a number of factors including the intensity of the acceleration and the period for which that acceleration is applied. Therefore, for convenience, accelerations are divided into long duration, ie more than one second, and short duration which essentially relates to impact acceleration forces.

Long Duration Acceleration

The human body is adapted to live under the influence of the force of gravity on Earth. Accelerations in aircraft can subject the body to forces of acceleration much greater than this, and for convenience are denoted as multiples of our 1G terrestrial environment. Such long duration accelerations are perceived as increases in body weight so that limbs become harder to move, the head becomes heavy, and organs are displaced from their normal positions. Since the increase in weight affects the blood, there are also changes in the circulation.

When sitting upright or standing the blood pressure measured in the upper arm closely reflects the pressure changes seen in the heart. However, there are hydrostatic forces that act on the blood pressure so that, at head level, the blood pressure is reduced from that seen at heart level. Similarly, the blood pressure in the lower body and legs is greater than that seen in the heart. Such hydrostatic differences are the result of the force of gravity and as applied G forces multiply the force of gravity, so too do they multiply the hydrostatic variation in blood pressure. This can have the effect of reducing the blood pressure at head level to such an extent as to stop all flow to firstly the eye and then the brain. Furthermore, such G forces drive blood down to the lower areas of the body, especially the legs, and make it very difficult for blood to return for recirculation. Blood pooling in the periphery exacerbates the hydrostatic consequences of applied G forces. For these reasons G forces cause greying out of vision and ultimately unconsciousness. This will occur at approximately 3.5G in the relaxed subject but the use of anti-G straining manoeuvres can (by increasing the blood pressure) delay the onset of loss of vision and consciousness until 7 or 8G, or perhaps even more. Such procedures are very tiring and are relatively short term.

G tolerance is reduced by a number of factors including hypoxia, hyperventilation, heat, low blood sugar, smoking, and alcohol. Fighter pilots are assisted in anti-G procedures by the inflation of anti-G trousers. These also help to reduce the fatigue associated with anti-G straining manoeuvres. Other devices designed to improve G tolerance include raising the legs of the pilot or reclining his whole seat but such systems are difficult to incorporate in existing aircraft and even when designed in from new are not a complete solution. The most important factors for the pilot are practice and experience.

Negative G occurs during aircraft manoeuvres such as inverted flight, outside loops, bunting, and some forms of spinning. The consequences on the body are the opposite of those for positive G but are even more uncomfortable. Symptoms include facial pain, bursting of small blood vessels in the face and eyes, the pushing up of the lower eyelid to cause 'redout' and the upward rush of blood from the lower body causes slowing of the heart. The maximum tolerable level is -3G, and then only for very short periods.

Short Duration Acceleration

The effects of impacts are directly related to the relative strengths of various parts of the body. The body can tolerate forces of, at most, 25G in the vertical axis and 45G in the fore to aft axis. Forces above these levels will cause injury or death. Therefore, in designing an aircraft it is as well to provide protection to ensure crash forces to not exceed these values.

Adequate restraint of the pilot in his seat is vital to prevent injury through contact with aircraft structures, and stop him being thrown out of the aircraft on impact. The best form of harness is a five point system with lap, shoulder and negative G (crutch) strap. The negative G strap holds the harness in place during inverted flight but moreover prevents the wearer from 'submarining' under the lap strap of the older, four point system, on forward impacts. Any harness system should be comfortable, easy to use and cause the minimum of restriction of normal movement. However it must be rigid under impact and resistant to inadvertent release.

1a.8 Anatomy and Physiology of the Ear

The structure of the ear was shown in Figure 1a.4. The ear fulfils two important functions: those of hearing, and balance. In hearing, sound that has passed down the external ear causes the eardrum to vibrate. These vibrations are passed across the middle ear by a series of small bones that act as an impedance matching device, ensuring that the vibrations of the eardrum are conditioned to be suitable to excite the fluid-filled part of the inner ear concerned with hearing – the cochlea. Problems of venting the middle ear have already been addressed, and the problems of noise damage and age on hearing are dealt with in 1b.1.

The other major function of the inner ear is in detecting angular and linear accelerations of the head. The structures that detect angular accelerations are the semicircular canals which may readily be observed in Figure 1a.4. Not apparent in this figure are the linear accelerometers – known as the otoliths – that are located at the base of the semicircular canals. Together, these angular and linear accelerometers are termed the vestibular apparatus. The function of the vestibular apparatus is to provide data to the brain that enable it both to maintain a model of spatial orientation, and to control other systems that need this information. For example, information from the vestibular system is used to control eye movements so that a stable picture of the world is maintained on the retina even when the head is moved.

There are two major problems associated with the vestibular system. The first is that it is not sufficiently reliable to maintain an accurate model of orientation without other – principally visual – information (see 2b.2), and the second is that motion stimuli detected by this apparatus can lead to nausea.

Motion Sickness

Motion sickness occurs when man is exposed to real or apparent motion of an unfamiliar form. It causes nausea, vomiting, hyperventilation, pallor, and cold sweating. It can be extremely incapacitating but is a normal response to the perceived stimuli. Anyone with a normal sense of balance will suffer motion sickness if provoked enough.

Motion sickness occurs at sea, in the air, in space, and may also be stimulated by the absence of motion when the sufferer is expecting movement such as during simulated aircraft flight. Although a few individuals suffer sickness on every flight many rapidly adapt to the motion and no longer suffer any symptoms. It is perhaps not surprising therefore that motion sickness is particularly a problem of flying training. However, if former sufferers are not exposed to such stimuli for a few weeks, such as during periods of leave or ground duties, a return to flying may be associated with a recurrence of symptoms until they become adapted again. Passengers are more prone to air sickness than crew for this reason. Incapacity from air sickness can be of great significance, however, to the parachutist, who may rush his departure from the aircraft and even mismanage his drills as a consequence.

There are a number of theories to explain why people react to motion stimuli by vomiting. It is certain the vestibular apparatus plays an important part in the detection of movement and the brain seems to compare signals coming to it from this system with those obtained from the eyes. A mismatch between the visual signals and the vestibular ones is a potent cause of symptoms. This mismatch may extend to take account of an apparent disparity between actual and expected stimuli. This would explain why it is that the experienced pilot is more prone to 'flight simulator sickness' than the novice who has yet to build a mental picture of the expected motion associated with the sensations of take-off, flight, and landing.

The incidence of motion sickness in flight varies greatly. It is influenced by the form and intensity of the stimulus, individual susceptibility, and the nature of the work being done. It occurs in a third of trainee aircrew, some of whom will be lost to flying training as a result. The provocative stimuli can be reduced by keeping the head still, as movement further aggravates the vestibular system. Visual mismatching can be reduced by closing the eyes, but for aircrew this is not an acceptable option. Nonetheless being relieved of look-out responsibility (and the associated head movement) and being made to concentrate on flying the aircraft may help the student greatly. Further familiarization before invoking more challenging manoeuvres such as aerobatics will allow the individual to adapt. There are medications that can alleviate the symptoms: the most commonly used is hyoscine. This can greatly reduce the symptoms but does have some detrimental effect on performance compared with the unmedicated (but sickness free) state. Other drugs are becoming available but in all cases the advice of a doctor experienced in aviation medicine should be sought lest a medication unsuitable for aircrew is prescribed.

1a.9 Vision

The effects of hypoxia and acceleration on the eye have been noted but a general understanding of the mechanisms of vision is useful. Of our five senses vision is the most commonly used and good vision for the aviator is essential.

Anatomy and Physiology of the Eye

The basic structure of the eye is shown in Figure 1a.5. The eye behaves like a camera with the lens of the eye focusing the rays of light onto the retina rather as the camera lens focuses light onto film. The retina converts light into nervous impulses that travel up the optic nerve to the brain where the visual picture is built up. The retina has two types of light sensitive cells called rods and cones. The rods are sensitive to lower levels of illumination than the cones, but are not sensitive to colour. Thus, at low levels of illumination, we can see only in monochrome. Even on a dimly lit flight deck, however, the colour coding of instruments must be visible and they must therefore be bright enough for cone vision to be used.

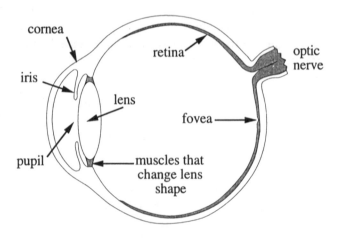

Figure 1a.5. The anatomy of the eye

The eye has two basically different ways of adjusting to different brightness levels. The first is by changing the diameter of the iris. This can change very rapidly to alter the amount of light falling on the retina by a ratio of about 1:5. Chemical changes in the retina are also used to adapt to different ambient luminances, and though these take place relatively slowly (full dark adaptation of the cones takes about seven minutes, with another 30 minutes required for full rod adaptation), they can cope with huge changes of luminance level (of around 150 000:1 just for the cones).

Despite this flexibility, it is possible for too much light to fall on the eye. Pilots, like Arabs and Eskimoes, are exposed to higher light levels than most humans.

Very high light levels prevail at altitude, and the light is reflected back at the pilot by cloud tops. This light comprises all colours of the spectrum, but contains more of the damaging blue and ultra violet (UV) wavelengths than are encountered on the surface of the earth. Blue light is very energetic and may cause cumulative damage to the retina over the course of a flying career. Prolonged exposure to UV can also cause damage – usually to the lens – but most UV is probably filtered by the cockpit transparency.

Wearing appropriate sunglasses can provide complete protection against these problems. The pilot is thus well advised to consult a knowledgeable supplier and request sunglasses that are impact resistant, that have thin metal frames (to minimize visual obstruction), that are constructed of hard-coated polycarbonate (for safety and strength), that are of good optical quality (refractive class 1), that have a luminance transmittance of 10–15 per cent, and that have appropriate filtration characteristics (meeting, for example, British Standard 2724 for special purpose sunglasses).

The capacity of the eye to resolve detail is termed 'acuity', and it is important to appreciate that acuity is not distributed evenly across the retina. The most

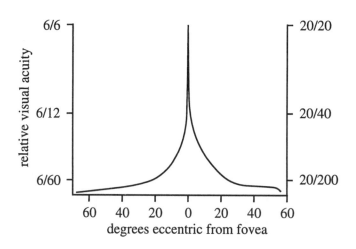

Figure 1a.6. The change in acuity across the retina

central part of the retina is termed the fovea, is composed only of cones, and is the part of the retina that you are now using to read this print. It is here that we may have 6/6 or 20/20 vision (the ability to see at 6m – or 20 feet – that which the so-called normal individual is capable of seeing at that range). Most pilots can resolve to a level of 6/5 or even 6/4 (they can see at 6m that which the average person would have to be at 4m to see). Although acuity does not drop with age, the ability to bring close objects into focus does. Although we have good resolving power at the fovea, this drops rapidly even a few degrees away from the fovea (see Figure 1a.6). Anything that must be interrogated in detail, and everything to which we attend, is automatically brought to fixation on the fovea. The rest of the retina fulfils the relatively coarse function of attracting our attention to movement and change, ensuring that we do not bump into things, and providing us with a sense of motion and orientation.

1b FLYING AND HEALTH

Introduction

This section addresses the problems that may render a pilot somewhat less than fit to fly. Since flying is a specialized occupation taking place in an unusual environment, and since the pilot may be responsible for the safety of many others, his own health and reliability are very important.

1b.1 Hearing Loss

In 1a.8 the anatomy and physiology of the ear were described. It will be seen that hearing depends on two mechanisms: an intact conductive system of eardrum and small bones (ossicles) to transmit vibration, and a transducer system, the cochlea. Any interference with the former will lead to a form of deafness known as 'conductive' and this may arise in early life from damage to the middle ear from infection or trauma. Such damage can also arise later in life but is often treatable with medication or surgery.

More serious for a pilot is a fault in the sound-receiving system, the cochlea, as this is usually irreversible. The sensitive membrane in the cochlea, which responds to vibrations and generates the nerve impulses which the brain interprets as sounds, can be damaged by overstimulation in the form of loud noise. The result is Noise Induced Hearing Loss (NIHL) and although this may be temporary at first, excessive exposure will lead to permanent damage and hearing loss. Noise levels are generally measured in terms of decibels (db) which relate to an arbitrary normal of 0 db. As a rough guide 0 db is the sort of sound (or lack of it) that is experienced in a sound-proofed room such as a recording studio, 30 db is like being in a library, 50 db like an average office,

70 db like a street corner, or average conversation, and 100–120 like a large jet aircraft taking off nearby. Amounts of noise energy in excess of 90 db for prolonged periods will cause damage – initially temporary lasting a few hours or days – but eventually permanent if the noise insult is prolonged. The damage risk is not simply related to length of exposure but is also affected by the total noise energy that is encountered. Thus, 90 db for eight hours will be as damaging as 103 db for half an hour, or 116 db for a minute.

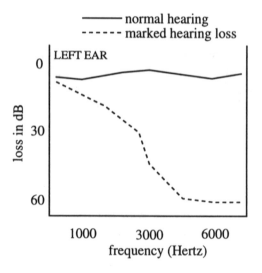

Figure 1b.1. Normal hearing and hearing loss demonstrated by audiometry

The importance of this is that any serious noise exposure can cause permanent damage to the hearing (such as illustrated in Figure 1b.1) and this does not relate exclusively to aircraft noise. The effects of gunfire – whether shotgun or rifle – as well as motor racing and less obvious sources such as discotheques and personal stereo sets at high volume can all be damaging and hearing protection should be used by those such as pilots whose job depends on retaining good hearing. Even simple insert ear-plugs of the soft wax variety are very effective in attenuating the damaging noise, but for those who wish to undertake noisy sports or hobbies on a regular basis a pair of good quality ear-defenders of the 'muff' type are essential.

Presbycusis

All pilots are going to suffer some hearing deterioration as part of the process of growing old; this is called presbycusis. Like noise exposure, the effects of ageing are to cut out the high tones first (see Figure 1b.2). Add this to a level of deafness already caused by noise exposure and this may be serious for a pilot who, as he reaches middle age, has to undergo regular hearing tests.

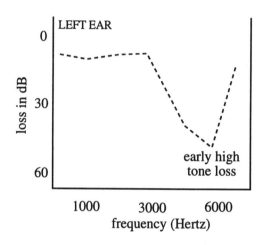

Figure 1b.2. Presbycusis: Early high tone loss, demonstrated by audiometry

1b.2 Visual Defects and their Correction

The mechanism by which the eye focuses light on to the light-sensitive retina was described in 1a.9. About 70 per cent of the refraction (or bending) of the light is produced by the cornea, which is fixed, and some 30 per cent by the lens, which can vary its focal length. Deficiencies in either of these can cause difficulty in obtaining a clear retinal image. In addition the shape of the eye can also affect visual acuity by determining exactly where the retinal image falls in relation to the focusing apparatus.

In long sightedness (hypermetropia) a shorter-than-normal eye results in the image being formed behind the retina and unless the combined refractive power of lens and cornea can accommodate to focus the image in the correct plane blurring of vision will result when looking at a relatively close object. A convex lens will overcome this refractive error.

In short sightedness (myopia) the problem is that the eye is longer than normal and this results in an image forming in front of the retina and accommodation by the lens cannot overcome this. Thus distant objects will be out of focus but close-up vision may be satisfactory. A concave lens will help to correct this.

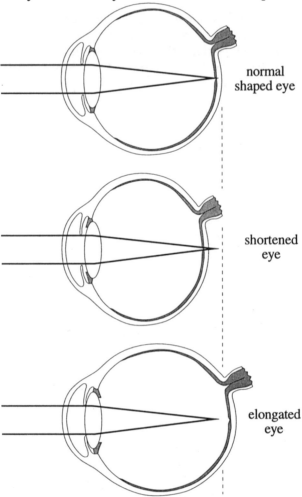

normal
shaped eye

shortened
eye

elongated
eye

Figure 1b.3. Visual defects

Pilots who need correction for either condition will usually be permitted to fly as long as their corrected vision allows them to read normal small print in good lighting at a distance of 30 centimetres, and to read 6/9 in each eye (which is often the next to bottom line on the standard optician's chart, read at a distance of six metres). This is equivalent to reading a car number-plate at about 40 metres, as opposed to the driving test requirement of 23 metres.

The ability of the lens of the eye to alter its focal length (accommodation) depends on its elasticity and this is gradually lost with age. Around the age of 45 the lens is unable to accommodate fully and a form of long-sightedness called presbyopia occurs. This normally starts with difficulty in reading small print in poor light especially when tired. Often a minor correction with a weak convex lens in the form of half or look-over glasses will suffice for this.

Pilots who need correction for both near and distant vision - for example, those who were myopic when young and who have become presbyopic - will normally be able to wear suitable correction with bifocal lenses but should seek expert advice from their Aviation Medical Examiner before choosing these. Likewise, contact lenses may be very convenient for some forms of visual defect in pilots but expert guidance should be obtained.

1b.3 Toxic Hazards

Aviation involves the use of many substances which are in themselves toxic or have the potential to become so in, say, a fire. The danger to aircrew is largely from inhalation of such substances. Possible sources of toxic fumes in an aircraft are the engine (exhaust gases, vapours from fuels, lubricants, and hydraulic fluids), the airframe (anti-icing, fire extinguishing agents), the cabin (products of combustion of upholstery, passengers' baggage), the cargo or payload (dangerous goods improperly packed, agricultural chemicals) and the atmosphere (ozone).

The most important of the exhaust gases is carbon monoxide (CO). This highly toxic odourless gas may enter the cabin when heat exchangers are used to provide the cabin heating, and accidents have been caused by faults in such a system. The effects are headache, breathlessness, impaired judgement, and eventual loss of consciousness.

Fuels and lubricants may produce irritant vapour and eventually drowsiness if inhaled.

Fire extinguishing agents may be toxic through suffocation, lung irritation and effects on the brain such as dizziness, confusion, and coma.

Ozone is a naturally occurring variant of oxygen which is irritant to the lungs in higher than normal concentrations. It is present in such levels at altitudes above 40 000 feet but most aircraft operating at these levels effectively break down the excess ozone at the compressor stage of ingested cabin air.

Agricultural chemicals include many very toxic insecticides. These can cause fatal poisoning at very low levels, and even traces can cause behavioural changes.

1b.4 Obesity, Diet, and Exercise

Obesity implies an excess of fatty tissue in the body but the exact definition is not clear. The Body Mass Index (BMI) relates height to weight by the formula BMI = weight (in kilogrammes) divided by height (in metres) squared. Thus a man of 180 cms weighing 75 kgs would have a BMI of 75 divided by 1.8 x 1.8 which equals 23.15. In general a BMI in excess of 25 in a man is regarded as above normal, and over 30 is obese. A person with a BMI in excess of 30 has a greatly increased risk of developing certain diseases (see below).

Although body weight (which usually means body fat, despite protestations about 'heavy bones') varies considerably between healthy people who take normal exercise the simple fact is that the ideal weight for an adult is generally about the weight that he was when he was 21. At this age the growth process will have been completed, the person will probably be quite active physically, and he will not have achieved an income level sufficient to allow him to eat to excess. This therefore is a good general rule of thumb when assessing a target weight when embarking on a diet to lose weight. There is no magic formula for this process. Quite simply the food taken in has to balance the energy expended in work and if there is an excess the extra will be stored as body fat. Since it is very difficult to use up extra food energy by exercise the only practical way to lose weight is to eat less, and this is the most effective way to avoid putting on weight. Plenty has been written about calorie controlled diets and there is no shortage of these. What has to be remembered is that although exercise is beneficial to general health it is a very inefficient way of burning off excess calories.

To reduce the risk of coronary artery disease exercise has to be regular, and sufficient to cause an increase in the pulse rate to something like double the resting level for at least 20 minutes, three times a week. Thus tennis, squash, jogging, or brisk walking are good for this, but golf, gentle swimming, or walking the dog are not necessarily very efficient.

Diseases associated with obesity are hypertension (high blood pressure), gout, diabetes, osteoarthritis (wear and tear on joints), and possibly coronary heart disease, although this latter may only be a risk if such things as blood pressure, diabetes, or gout are involved.

1b.5 Coronary Artery Disease, the ECG, and Raised Blood Pressure

The heart is a muscular pump sending blood around the body and its muscle requires its own supply of blood to provide it with food and oxygen. The arteries that carry this are the coronary arteries and with increasing age these often become narrowed or even totally blocked. Narrowing or gradual blockage results in insufficient blood reaching the muscle and the effect is to deprive part of the muscular pump of oxygen when demands are placed on it by exertion or emotion. The result is pain in the chest, or sometimes the neck or the shoulders or arms, especially on the left side. This pain, developing with effort and often subsiding on resting, is called angina and indicates serious impairment of coronary blood flow. This may lead to inefficient action of the heart and gradual, or occasionally sudden, heart failure may result.

If a coronary artery suddenly blocks (usually due to a clot or 'thrombus' forming in an already compromised vessel - hence 'coronary thrombosis') the effects are more dramatic, often with sudden severe chest pain, collapse, and sometimes complete stopping of the heart. This cardiac arrest is rapidly fatal if not reversed by some form of resuscitation. Even if this extreme stage is avoided a sudden blockage with interruption of blood supply leads to the death of an area of heart muscle - so-called 'infarction' - so the other term used for such an attack is myocardial infarction.

It is obvious that a person at risk to such an episode is not fit to hold a pilot's licence. There are no simple accurate tests that will give a picture of the state of the coronary arteries but the ECG (electrocardiogram) is of some limited value. This measures the electrical activity in the heart as the muscle is working and abnormalities may indicate coronary artery narrowing. Sometimes the ECG shows that an infarct has occurred months or even years beforehand and has either been totally symptomless at the time, or symptoms have been thought of as indigestion or some similar minor problem. Even in an apparently healthy person, evidence of previous infarction means that they are at greater risk to another attack than a person with a normal ECG.

Risk factors for developing coronary heart disease are family history, smoking, raised blood pressure, raised blood cholesterol, lack of exercise, and diabetes. Other things such as stress, obesity, alcohol, and certain dietary variations are less clearly understood.

All of these, apart from the genetic background, can be remedied. In particular the regular monitoring of blood pressure makes it possible to arrange early treatment to control this and with modern drugs a pilot may continue to fly after appropriate checks. Since raised blood pressure is also the main risk factor for the development of a stroke this also can be influenced.

1b.6 Incapacitation in Flight

Dramatic sudden incapacitation of a pilot is extremely uncommon and very rarely the cause of an aircraft accident. Gradual or insidious incapacitation may be unnoticed by the pilot or other crew and is probably under-reported as it usually happens at a stable phase of flight and is often short-lived.

The purpose of the rigorous medical selection process for pilots and their periodic health checks (increasingly frequent as they get older) is to try to minimize the risk of sudden incapacitation and it is at least partially vindicated by the rarity of such events, especially in relation to heart disease and epilepsy. The commonest cause of sudden incapacitation is acute gastroenteritis, but this is usually easily managed and the incapacity is rarely total. Of the more serious forms a 'fit' or 'faint' in an apparently healthy person is the commonest. Sometimes further enquiry reveals that the attack is due to alcohol withdrawal and could have been anticipated. Renal colic – a severe pain caused by a kidney stone – or gall bladder colic have been reported and are difficult to predict. Loss of consciousness from heart disease, either from an acute disturbance of rhythm or a sudden episode of coronary heart disease is extremely rare.

With the advent of high quality simulators incapacitation training has become feasible and the mandatory training now required has proved so effective that pilots with a known higher-than-average risk of sudden illness can be allowed to fly in a multicrew operation.

1b.7 Fits and Faints

Any predisposition to sudden loss of consciousness is clearly not acceptable in a pilot and a history of such an episode requires careful investigation before the granting, or restoration, of a flying licence. The term fit (or 'seizure') is usually reserved for some manifestation of epilepsy, whereas faint (or 'syncope') refers to a change of consciousness caused by a disturbance of blood flow to the brain.

Epilepsy is not a specific disease but refers to a set of signs and symptoms which occur in response to some disorder of electrical activity in the brain. It is often described in terms of major or minor epilepsy although the distinction is not always clear. Thus major fits are usually accompanied by convulsions or other uncontrolled movements while minor fits are often no more than short periods of a few seconds of 'absence' or loss of attention. Both, however, are associated with distinct loss of consciousness and are therefore an absolute bar to a flying licence. In most patients these conditions would manifest themselves at an early age and they would be excluded from training as pilots but in some instances the condition may arise in later life as the result of a head injury, or perhaps other disease process, and a pilot may find himself permanently grounded as a result. Even when seizures are well controlled with drugs and a person has regained a driving licence, no form of flying licence would be granted.

In many patients with epilepsy the EEG (electroencephalogram) is abnormal. This test consists of a recording of the very small voltages produced by the brain during its routine activity and is carried out by placing electrodes on the scalp to detect these. Certain characteristic abnormalities of the waves so produced can often be diagnostic and the test will be applied routinely to pilots who are being investigated for an episode of disturbance of consciousness. It is sometimes also used in initial assessment of pilots.

Faints or syncopal attacks are more commonly a cause of loss of consciousness in adults who have no previous history of blackouts. Usually the cause is fairly simple. An otherwise healthy person may faint from shock, loss of blood, lack of food or fluid, or other physiological stresses, and this sort of attack has no sinister significance as far as future flying is concerned as long as the cause is clearly defined. In all these instances the basic mechanism of the attack is a sudden reduction of blood supply to the brain and a common cause is standing up quickly after prolonged sitting, especially when hot or dehydrated. As long as the cause can be readily identified and other abnormalities can be excluded a pilot will be allowed to return to flying, albeit perhaps with a restriction to fly with another qualified pilot at first.

1b.8 Psychiatric Disease, Alcohol, and Drugs

Serious forms of psychiatric illness associated with loss of insight or contact with reality - a psychosis such schizophrenia or manic depression - will result in the permanent denial of a flying licence. Less serious illnesses of the neurotic type such as anxiety states, phobic states, obsessional disorders, and depression are all a bar to flying while they are active, or under treatment, even though the symptoms may be well controlled. The use of any medication affecting the brain is clearly incompatible with safe flying. Most pilots who suffer such illnesses, however, will return to flying after a suitable period of good health off all drug treatment.

Alcoholism is not always easily recognized or defined. A World Health Organisation definition refers to it as being present when the excessive use of alcohol repeatedly damages a person's physical, mental, or social life. The single most important characteristic of the alcoholic's use of drink is a loss of control either in a chronic form, with an inexorable progression to more and more destructive patterns of drinking, or as exhibited by regular binges of uncontrolled drinking lasting days or even weeks.

Very few alcoholics, at least in the early stages of the condition, present with the picture of physical, mental and moral decay of popular literature. Most are (apparently) 'sober citizens'. They have responsible jobs such as doctors, managers, craftsmen, pilots and the like. In fact no occupational group is exempt from the illness but some - such as aircrew - have a higher than average risk because they are exposed to more of the things known to be associated with its development. These include social isolation, boredom, high income, easy access to cheap alcohol, a 'drinking culture' with a need to conform and be gregarious, and the frequent use of alcohol to unwind and perhaps aid sleep.

A high social level of alcohol intake can be damaging even without dependence developing. Damaging levels (ie those that will cause some physical damage to liver, heart, brain, blood cells, or other organs in 50 per cent of a population) are surprisingly low. For men, six units daily or more than 30 per week is believed to be sufficient, where a unit is equivalent to a half pint of beer, or a standard glass of wine or spirit. For women the levels are probably a third lower, ie four units a day or 20 per week. Note that these are the DAMAGING levels for half the population. There is not complete agreement on what the SAFE limits are, but these are probably in the region of 21 to 28 units per week for men and 14 to 21 units per week for women.

Other questions that are often asked relate to the rate at which alcohol is excreted from the body and to any measures that can be adopted to speed the process. To answer the second question first, only time influences with any certainty the rate at which alcohol is eliminated from the body. Exercise, black coffee, food, and herbal remedies have no effect. Food can have some influence on the rate of absorption - ie alcohol taken on a full stomach will be more slowly absorbed than on an empty one - but this does not affect the rate of elimination of the alcohol once in the blood. The only thing that can affect this is time, and even then the rate varies. A rough guide - and it must be stressed that it is so rough as to be unreliable - is that it takes one hour for each unit of alcohol to leave the body, starting to count at least one hour after commencing drinking. Thus, six pints (12 units) of normal beer taken between 1800h and 2300h will not leave the body free of alcohol until 0700h the next morning.

The well known 'eight hours bottle-to-throttle' rule must be interpreted with this in mind: in a recent case a bus driver was found to be over the legal limit for driving the morning after drinking the equivalent of seven pints of beer, having had a good sleep and a meal in the intervening 12 hours. Internationally, legislation is increasingly being introduced to limit the maximum level of blood alcohol for piloting an aircraft. Since, however, even the smallest amount of alcohol can have adverse effects on performance it follows that every precaution should be taken by pilots to ensure that they are alcohol free when operating.

Danger signs of early problems with alcohol control are regular drinking alone, gulping the first drink, having to increase the intake to feel good, memory loss for the night before's events, morning shakes and other withdrawal effects, feelings of guilt about drinking and anger if criticized, and any adverse effects at all on family, work, or other social life.

Suspicion of alcohol excess in a colleague requires a prompt, frank, and positive approach with the knowledge that help is available. Given suitable treatment a pilot can return to flying, but total abstinence is the only realistic goal as it is the only sure end point. There is no hope of a return to controlled drinking for a pilot who has once had a serious problem.

Drug dependence requires specialist help which again could result in a pilot's returning to flying if successful. Any use of illicit drugs is incompatible with air safety and there is no place for even so–called soft drugs: they all affect performance, mood, and health.

1b.9 Tropical Diseases and their Prophylaxis

Very few tropical diseases are limited strictly by geography and many were prevalent in Britain until quite recently. Malaria was present in marshy parts of the Thames estuary this century, and cholera killed many in England in the last. Most are better regarded as diseases of poor hygiene and sanitation as this gives the best understanding of how to avoid them.

Malaria is still one of the world's biggest killers. It is transmitted by mosquitoes and the better managed parts of the world start by eradicating the marshy areas which are the insects' breeding grounds. Avoiding such areas, and avoiding being bitten by the wearing of long sleeves and trousers in the evenings when the mosquitoes are active, using insect repellant creams, sleeping in air conditioned rooms or under mosquito nets after spraying the room with insecticide, are all sensible but often neglected precautions. Too much reliance has been placed on anti-malarial tablets for these are never a complete protection as strains of the malaria parasite develop resistance to the drugs. Some airlines and other organisations employing international travellers are now moving away from the routine use of anti-malarial drugs and, instead, relying on a personal issue of a treatment pack containing a drug, such as Halofantrine (Halfan), that the person takes if he develops any suspicious signs or symptoms. As these are often similar to influenza, it is inevitable that sometimes a person will take the malaria cure when suffering only from 'flu, but this is felt to be a safer system than relying on possibly ineffective prophylaxis. There may, however, be some side effects from the drug used that makes it necessary for the pilot to be off all flying duties for a short period.

Yellow fever is insect borne but the primary protection against this is vaccination. The vaccine is extremely efficient, very safe, and effective for at least ten years. It is so good that it remains the only vaccination that can be demanded as a condition of entry to certain countries, and therefore is usually needed by international pilots.

Other vaccines are useful in varying degrees but none gives the almost certain protection of yellow fever vaccination. Polio vaccine is very effective, and typhoid vaccine (whether by mouth or injection) gives over 70% protection. Gammaglobulin gives limited and short-lived protection against hepatitis-A, but new active vaccines (eg 'Havrix') are becoming available that give greater and more prolonged protection: their current disadvantage is their high cost, but this would become economically sensible for stays in excess of a few weeks or where the maximum possible protection is required. Cholera vaccine is virtually useless in providing protection to the individual although it may still have some value in preventing epidemic spread of the disease. There is now no official support from the World Health Organisation for mandatory cholera vaccination of travellers and most airlines have ceased to use it routinely for their crews.

All these diseases - and many others such as dysentery, salmonella food poisoning, and tapeworm infestation - are food or water borne and are often the result of unhygienic food preparation, storage, or serving. Where hygiene standards are poor, these diseases can largely be avoided by some simple precautions. Only bottled water should be used, and ice for drinks made from water of similar purity. Salads and unpeeled fruit should be avoided unless adequately washed in purified water, and only hot, freshly cooked food that has not been standing around for hours (or days) should be eaten. Foods that are particularly risky in areas of poor hygiene are cream, ice cream, and shellfish.

All travellers should be aware of the higher incidence of any of the sexually transmitted diseases in less developed countries. Although AIDS has had all the publicity, the means of acquiring it does not differ from any of the other diseases, and likewise the means of avoiding it.

1b.10 Common Ailments and Fitness to Fly

Even minor degrees of ill-health can cause deterioration of flying performance, and the decision whether or not he is fit to fly requires careful thought by a pilot. In general, given the difficulty in making this subjective assessment, it is always better to take the line that any illness sufficient to raise doubts about fitness is probably serious enough to justify cancelling a duty.

The Common Cold, especially if associated with fever as it often is in the early stages, is a good example. Most people will recognize the lethargy, difficulty in concentration, and general malaise felt at the beginning of such an illness and these effects can seriously impair performance. Also associated is congestion of the upper air passages and this can lead to difficulty in equalizing pressure in either the sinuses or the middle ear spaces when in conditions of changing barometric pressure. Severe pain can result from the condition called Barotrauma and this can be incapacitating. In addition, even in the absence of serious pain, a blockage to the middle ear can lead to temporary deafness.

Gastroenteritis, usually from food poisoning, is common in travellers. Nausea and vomiting, diarrhoea, abdominal cramp-like pains, and fever are all common symptoms of this condition and a pilot is unfit to fly while affected. Most attacks are of the short-lived non-specific type known generically as 'Travellers' diarrhoea' and will settle spontaneously in two or three days at the most. Treatment with a gut sedative such as Loperamide (Imodium) may control the symptoms until spontaneous recovery occurs, but this does not mean that a pilot so afflicted and medicated is fit to work. Gastroenteritis which does not settle, with or without treatment, in 72 hours needs further investigation as the cause may be a more serious infection such as Salmonella.

Any medication with drugs, whether prescribed or self-purchased, is almost certainly a reason for not flying. A condition whose symptoms are bad enough to need medication could affect performance, and no drugs exist that are completely free from side effects. Examples are drowsiness with anti-histamines which are a common ingredient in proprietary cold cures and hay fever remedies, gastric bleeding with aspirin, and blurring of vision with drugs that stop bowel spasms in gastroenteritis. In general a pilot should seek informed aviation medical opinion before deciding to fly while taking any medication.

Two final points should be noted. First, that new drugs are constantly becoming available, and thus the pilot should consult an informed doctor or specialist travel clinic if he wishes to receive the most up to date advice on the prophylaxis and treatment of illnesses associated with travelling abroad. Second, anybody who develops any illness, especially a fever, on return from abroad should seek prompt medical advice and inform the doctor of the locations that have been visited.

PART II

BASIC

AVIATION

PSYCHOLOGY

2a HUMAN INFORMATION PROCESSING

Introduction

The task of flying an aircraft involves observing and reacting to events that take place within the cockpit and in the environment outside the aircraft. The pilot is required to use the information that he senses in order to make the decisions and take the actions which will ensure the safe path of the aircraft at all times.

The aim of this chapter is to enable pilots to develop an awareness of the system by which we receive and process information in order to make decisions, together with the implications of this system for pilot workload assessment.

2a.1 Basic Plan of Human Information Processing

Since the system by which we process information once it has been received by our senses cannot be described by relating each of the stages to distinct physiological structures in the brain, functional models have been developed to represent the mechanisms by which information is received, decisions are taken, and responses are selected and executed. We can use these models when errors are made to determine whether they have arisen as a consequence of a failure of perception, as a consequence of a failure of memory, or as a consequence of having successfully interpreted the information but having failed to take the appropriate action. Information processing models also enable us to understand better other factors, such as stress, that may influence human performance.

All information processing models are based on the assumption of a series of stages or mental operations which occur between information being received by the senses and responses being made. Figure 2a.1 illustrates a typical model. The way in which these stages influence our ability to use the information around us, to make decisions, and to perform actions is described in the following sections.

2a.2 Sensation and Sensory Memory

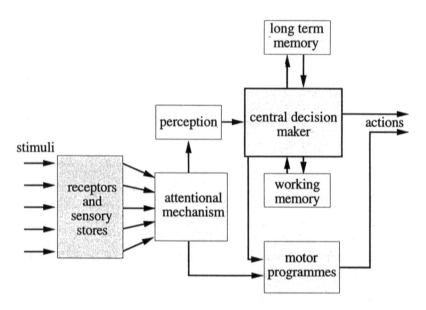

Figure 2a.1. A functional model of human information processing

Physical stimuli in the form of sounds, visual patterns, etc are received by the sensory receptors (eg eyes, ears) and stored for a brief period of time after the input has been terminated.

The key features of this system are:

a) that the information is physically represented at this stage, ie in the form of sounds or shapes,

b) that there is a separate store for each sensory system,

c) that the input decays rapidly.

In fact, information in the visual sensory store (iconic memory) lasts for between 0.5 and one second, and information in the auditory sensory store (echoic memory), for between two and eight seconds. The importance of sensory memories is that they enable us to retain information for a brief period of time until we have sufficient spare processing capacity to deal with the new input.

It is possible, for example, to hear a clock chiming, but to realize that you want to know the time only after the second or third stroke. In such circumstances echoic memory can be interrogated or 'replayed' to enable the strokes to be counted consciously. Similar effects can prevail with RT messages if you realize that a message is intended for you only when some of the message has already been passed.

2a.3 Perception and Mental Models

Perception involves the conversion of the sensory information into meaningful structures; eg a pattern of sounds becomes recognized as a particular message. The 'percept' is not a complete representation of the information in the sensory store, but an immediate interpretation of it. The process is complex as the physical information is coded using concepts (or descriptive labels) already existing in memory. It is thus true that we can 'perceive' only that which we can 'conceive'. The process is also influenced by the amount of processing capacity that the person is able to give to the information arriving at the sensory store and the nature of the information being received. Since channel capacity is limited in this way, much of the information that is 'sensed' is lost without being 'perceived' (see 2a.9).

The purpose of our perceptual processes is to create an internal model of the outside world. This model is plainly based heavily on the information provided by our senses, but not entirely so. As noted above, our experience and expectations of the world are also very important in the creation of the mental model. For example, aircraft RT conversation is notoriously incomprehensible to the layman but the pilot is able to perceive (ie generate a mental model) of it because he is both experienced at dealing with the physical constraints and distortions of the auditory signal and, more importantly, because he has strong expectations with regard to the potential content of the information. The system does however contain inherent dangers in that having developed a mental model we tend to seek information which will confirm the model and ignore other sources of input. The implications of this process for the pilot will be given further consideration in 2b.1.

The most important point to bear in mind about perception is that we do not perceive the world in a completely deterministic way. Our percept or mental model is based both on the information sensed by our receptors and on our expectations of the world. A good example of this is provided by the judgement of range or distance.

There are many cues available in the visual system to enable our mental model to be three-dimensional or include an impression of depth. These include convergence (the amount that our eyes converge in order to bring a visual target on to each fovea), stereopsis (near objects produce images on each retina that are more different from one another than distant objects), obscuration (near objects occlude distant objects), atmospheric perspective (objects become less coloured and less distinct as they become more distant), and retinal size (objects become smaller on the retina with range).

The last of these cues - retinal size - is of particular importance to the pilot. For example, in the late stages of an approach, he is likely to judge his height above the ground from the retinal size of the runway. In order to use this information, however, he must have a stored knowledge or expectation of the likely size of the runway. Thus if the runway is narrower than he expects it to be, then he may overestimate his height, and vice versa.

2a.4 Central Decision Making and Response Selection

Once information has been perceived a decision must be made about what should be done with it. It may be used to initiate an immediate response or be entered into part of the memory process. For example, on hearing a warning sound the operator may switch off the system, in which case the decision will involve the selection of a response. Alternatively, the operator may decide to hold the information in memory whilst a search is made for the problem which has triggered the warning. Information may be continuously entered and recalled from memory in order to contribute to the decision process. However, at this point a decision must be made either to retain the information for a short time in working memory by actively rehearsing (or internally repeating) it, or to attempt to learn the information and store it permanently in long term memory.

The decision making process and the use of judgement is central to human performance and is dealt with in 2b.8.

2a.5 Mechanisms and Limitations of Working Memory

Working memory is also referred to as short term memory. This memory system enables information to be retained for a short period of time; for example for the period between hearing a frequency on the RT and selecting it on a radio box. Verbal information is normally maintained in an acoustic form and spatial information using a visual code. Since verbal information appears to be stored in some sort of acoustic code, errors in verbal working memory normally take the form of acoustic confusions (eg 'cat' may be recalled instead of 'mat'), and this is likely to occur even if the verbal material is presented in written form.

Information is maintained in working memory by the process of rehearsal, with acoustic information generally being considered to be easier to retain than visual information because it is easier to rehearse sounds than information in a visual form.

The capacity of working memory is fairly limited in that the maximum number of unrelated items which can be maintained when full attention is devoted to rehearsal is about seven. This has important implications for the design of check-lists and for the amount of information that should be included in any single RT message. Once this limit of about seven is exceeded one or more items are likely to be lost or transposed.

The number of items retained in working memory can be expanded by the clustering or 'chunking' of related material. This can occur when two or more items have been previously associated, for example a familiar telephone code such as 071 may be clustered and require only one space rather than three. Chunking unrelated items into groups of three and four, eg 055 2781 will also assist the maintenance of items. The size of a 'chunk' will clearly be determined by the individual's familiarity with the information.

Unless actively rehearsed, the information in working memory will be lost in 10 to 20 seconds. The information in working memory is lost by the process of interference, which causes the information to become confused or replaced by that which was previously stored or the arrival of new information. The impact of interference and hence forgetting may be reduced by increasing the time between the arrival of inputs to be held in working memory and by reducing the similarity between the items.

2a.6 Mechanisms and Limitations of Long Term Memory

The information retained in long term memory can be classified into two types.

Semantic Memory

This includes the knowledge which we have which is associated with the things we are able to do, eg understanding a word, knowing the items on a checklist. This memory is our memory for meaning. The organization of the enormous amount of information which we all hold in semantic memory is the subject of much interest since it is known that our ability to use information presented to us, for example on a computer data base, will be greatly enhanced if the knowledge base is organized in a similar way to our semantic memory. It is generally thought that once information has successfully entered semantic memory the information is never lost. When we are unable to remember an item, eg a word, it is because we are unable to find where the item is stored in the memory system, not because the word has been lost from the store.

Episodic Memory

This memory system includes our knowledge about specific events. For example our memory for a particular flight or incident on the flight deck is held in episodic memory. One of the important features of episodic memory is that the information stored does not remain static but is heavily influenced by our expectations of what *should* have happened. Our recollections from episodic memory are thus influenced by our expectations of the world in a similar way as are our initial perceptions. This tendency for us to remember what must have been rather than what was is obviously problematic for accident investigators, and can be especially so for expert witnesses. A pilot observing an aviation accident will have much stronger expectations of a likely set of events than a lay observer, and his recollections may thus conform much more closely to his interpretation of his observations than the lay observer's recollection of the raw events.

Lastly, it should be noted that most common amnesias affect episodic but not semantic memory. Shock and physical damage may well prevent recall of events, but are most unlikely to affect the store of basic knowledge of the world.

2a.7 Motor Memory (Skills)

On many tasks, practice leads to a reduction in the amount of central processing capacity required and may eventually lead to the execution of the task becoming automatic. For instance, in learning to drive a car, the initially demanding task of changing gear can eventually be performed with little or no awareness. The process of skill acquisition is often divided into three phases. In the first 'cognitive' phase, the learner must think consciously about his actions and what he wishes to do. In the second 'associative' phase, the separate components of the overall action become, with practice, integrated until in the third 'automatic' phase the action can be executed as a smooth unit without conscious control. All that is normally required is an initial conscious decision to initiate the skill.

In skills such as driving a car or flying an aeroplane these behavioural sub-routines or 'motor programmes' may not require continuous conscious control, but they do require conscious monitoring. Such motor programmes can be very complex and enable many actions to be carried out simultaneously (consider, for example, the control required by an organist using both hands and feet) while the central decision maker is otherwise occupied. It is easily possible to fly (using motor programmes) and to maintain a conversation (using the central decision maker) at the same time, but if the flying becomes difficult, then the central decision maker has to be devoted to it and the conversation stops. It thus appears that when we say we can do only one thing at a time, we mean that we can do only one thing that requires central processing capacity at a time - automatised actions or motor programmes can run concurrently. Unfortunately, although the advantage of the development of motor programmes is that they free our processing resources for other activities, they have the disadvantage of being open to particular types of error. These errors (sometimes referred to as action slips) tend to occur at the initiation stage rather than the execution stage of action. Thus an instance of a typical error would be the business executive who lifted the telephone and shouted 'come in'. Since actions controlled by motor programmes are performed without conscious awareness, it can be a considerable period of time before the individual realizes that his behaviour is not as planned. The implications of this for errors on the flight-deck is considered in 2b.6.

2.a.8 Response Execution and Feedback

At the response execution stage a decision is translated into a specific sequence of motor commands. The responses will be in the form of conscious actions or words or both, unless they are being driven by motor programmes.

In a situation where there is pressure to make a rapid response, as can happen in aviation, several factors must be borne in mind since there will frequently be a trade-off between speed and accuracy:

a) The obvious fact that in certain situations where any delay can have catastrophic consequences there will be pressure to make a response before sufficient information has been processed.

b) Conditions which increase our arousal level will lead to faster but less accurate responding.

c) Since auditory stimuli are more likely to attract our attention than visual stimuli, they are more likely to be responded to in error.

d) If we expect a stimulus and prepare the appropriate response, we will respond more rapidly if the expected stimulus occurs. If however, an unexpected stimulus occurs we will be more likely under pressure to make the prepared response, ie to make an error of commission.

e) An increase in age between 20 and 60 years tends to be associated with slower but more accurate responding.

When flying an aircraft or performing any skilled task we continuously monitor both the environment and the consequences of our action. For example when flying an aircraft, we make use of the closed-loop feedback system between the response outputs and sensory inputs.

It must be emphasized that the detection and interpretation of sensory inputs will be greatly influenced by what we are expecting to receive. Thus decisions which have already been made or information stored in either short term working memory or long term memory may influence the perception or pattern recognition of the input.

Response Times

Response or reaction time (RT) is the time between the onset of a given signal and the production of a response to that signal. In the simplest case, we can imagine a subject resting his finger on a button, looking at a light and waiting for it to illuminate. In this instance, the RT is likely to be about one fifth of a second (200msec). If, however, the subject is provided with two lights and two buttons, and is required to press the left button when the left light illuminates, and vice versa, the RT will be materially extended, and we can attribute the extension to the increased decision making load imposed by the task.

In flying, RTs will be even longer since the decision element of the task is likely to be even more complex. In the Boeing 737 that caught fire at Manchester(1985), the crew heard a bang well before V1, and the captain shouted 'Stop' within one second. This RT can be considered fast for a real life situation, and was enabled to be so since the situation was relatively clearly defined. In the HS748 that ran off the end of the runway at Sumburgh (1979) with elevator locks engaged, however, there were some five seconds between the call 'Rotate', and the application of the brakes. Subsequent simulations showed this time to be about average for a number of pilots. The time here was clearly extended because the pilot was not expecting the controls to be locked (having just checked them), had not trained for the eventuality, and was expecting to deal with any problem that arose at that juncture in the air.

Although there are instances in flying, such as the two mentioned above, in which RT can be important, pilots are generally advised that it is of paramount importance to make the correct rather than the fast response.

2a.9 Attention and the Limitations of the Central Decision Maker

There are two potentially limiting stages to the processing of information. One is the limit to the number of items which can be maintained in working memory. The other limitation concerns the rate at which information can pass through the system - the 'channel capacity'.

Although it is difficult, perhaps impossible, to describe in precise quantitative terms the channel capacity of the system, the fact that it is limited means that we are not able to devote conscious thought or 'attend' to all of the stimuli that impinge upon us. Thus some form of mechanism is required at an early stage of the system to allow us to select the stimuli that will be perceived consciously and used as the basis for thought and decisions.

Two types of attention are sometimes described.

Selective Attention

The term 'selective attention' is used to describe the process by which inputs are sampled to ensure that the information which receives detailed processing is that relevant to the task in hand. It would be unfortunate, however, if this process were so rigid that information not directly related to the subject occupying the central decision maker was always lost, since this would mean that if we were concentrating on something we would fail to hear someone calling our name. In fact, people are rather good at detecting information of relevance to them, such as their name or aircraft call sign, even if it is presented on a non-attended input channel. This phenomenon is often referred to as the 'cocktail party' effect and is interesting in demonstrating that the information in a non-attended channel (of which we have no conscious awareness) must be receiving a good deal of analysis in order that it can be tested for salience or importance.

Apart from our name and call sign, some stimuli such as loud noises and flashing red lights are particularly 'attention-getting' and are therefore important in the design of warnings.

Divided Attention

We are not always able to give our central decision making channel over to one task on a continuous basis. The pilot flying a visual approach will be required to divide his attention between looking ahead outside the aircraft and looking down at the airspeed indicator. Even when one task is being carried out by a motor programme it is beneficial if the central processor is occasionally diverted from its primary occupation in order to check and monitor the progress of the motor programme. If, however, the central processor becomes so consumed or preoccupied with its main task, then the progress of the motor programme can go unchecked and lead to error (see 2b.6).

The account given above of a single decision channel model of human information processing is necessarily simplistic and much more sophisticated models have been developed, but they generally share the ideas of a central decision making channel with subordinate activities carried out more automatically, and they do help to explain and understand many of the information processing activities carried out by pilots.

The Influence of Stress on Attention

The usual effect of stress is to increase the arousal level or general activation of an individual. The level of arousal influences the scanning pattern and hence the perception of information by an individual. For instance, a pilot during a quiet period of the cruise phase of flight, when his arousal level may be low, does not scan the instruments as frequently as he does during the approach and landing phases when his arousal will be at a significantly higher level. Under conditions of high stress and arousal, the sampling rate may be increased, but the pattern of sampling is reduced to a narrower range of stimuli as a consequence of attention being restricted to the primary task (sometimes termed 'narrowing of attention'). This can lead to the undesirable situation of important information required in an emergency being missed by a pilot because his stress response to the situation caused his attention to be restricted to the primary source of the problem. The influence of stress on performance is described in 3a.2.

2a.10 Mental Workload

Since our ability to process information is limited, this has obvious implications for the level of performance we are able to achieve. There are various definitions of the term 'workload', but all involve some relationship between the imposed demands of the task and the availability of channel capacity or mental resources to deal with those commands. Workload assessment in the field of aviation arose from the need to ensure that the demands of flying the aircraft do not exceed the capabilities of the pilot. Having said that, it may be that some pilots (for example, the single seat fast jet pilot) may work at maximum workload for considerable periods, simply prioritising the tasks with which they are confronted to match available channel capacity.

The subjective experience of workload arises from the requirements of the task, the circumstances under which it is performed (especially the time available to perform it), and the skills, behaviours, perceptions and subjective state of the operator (particularly if he is under any stress).

The relationship between workload and performance can be conceived in the form of an inverted 'U' curve. Human performance at low levels of workload is not particularly good. If we do not have enough to do we quickly become bored and tend to lose interest in the task we are required to perform. Thus at low levels of workload (underload), errors typically take the form of missed information due to lack of samples of the input from the environment around us, eg the aircraft instruments. There is some concern that the lack of workload during the cruise phase of long haul flight could potentially affect the ability of pilots to react quickly in an emergency.

As the demands of the task, or the workload, are increased, the standard of our performance increases until an optimum level of workload and performance is achieved. Any increase in workload after this point leads to a degradation in performance. At extremely high levels of workload (overload), important information may be missed due to the narrowing or focusing of attention onto only one aspect of the task.

The human information processing model can help us to determine the source of overload. It may be that the task is too difficult, ie the amount of information to be perceived in order for a decision to be made is beyond the attentional capacity of the pilot (called qualitative overload), or alternatively there may be too many responses to be made within the time available (called quantitative overload).

Within the aviation industry the concept of workload is of primary importance. It is used as a predictive tool to ensure that, when a new system is being developed, some assessment is made to ensure that the pilot will have sufficient spare information processing capacity to cope with failures or unexpected events. Workload assessment techniques may also be used to identify and to assess the source of temporary bottlenecks which may occur in the system when the demands of the task exceed the capabilities of the pilot and performance deteriorates. Alternatively, workload assessment techniques may be used to compare two alternative pieces of equipment or to enable a comparison to be made between the performance of individuals. For example, it has been shown that flight instructors differ from their students in their level of residual spare capacity. Within aviation workload techniques are also employed for:

a) the evaluation of new cockpits, changes to the systems or operating procedures.

b) aircraft certification.

c) the determination of crew complement.

2b COGNITION IN AVIATION

Introduction

In this section, the basic outline of human information processing provided in chapter 2a is applied to the aviation context, and some examples of information processing difficulties that may be encountered in aviation are provided.

2b.1 Erroneous Mental Models

Mental models are usually accurate and bear a good relationship with the real world, but sometimes the data provided by the sensory systems can lead to a model being made which is inaccurate. When a mental model differs from the real world this is known as an illusion. Figure 2b.1 provides an example of a visual illusion: the distances AB and BC are equal on the page, but unequal in the mental model.

Visual illusions in flying are frequently associated with inappropriate experience or expectation on the part of the pilot. For example, the pilot who is accustomed to flying in relatively polluted air may have learned to use 'atmospheric perspective' (the tendency for objects to become indistinct with distance) as a good cue to range. If he operates in a very clear atmosphere, he may then perceive distant objects as being closer than they are. Similarly, if a pilot is expecting to see a large distant object, he may mistake this for a similar, but smaller and closer object. This effect may well have played a part in the Air New Zealand DC10 accident in Antarctica (1979), and enabled the crew to mistake small, close cliffs for the large distant cliffs that they would have been expecting to see.

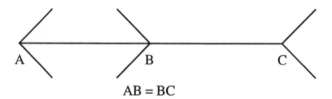

AB = BC

Figure 2b.1. The Müller-Lyer illusion

2b.2 Spatial Disorientation

There are a number of ways in which the pilot can suffer from illusions of orientation. Some of these are caused by the misinterpretation of visual information (vision is the most important contributor to the perception of orientation). For example, a pilot may interpret a sloping cloud bank as a level horizon, or even misidentify ground lights (perhaps a flotilla of fishing boats in a featureless sea) as stars, and make a correspondingly inaccurate mental model.

Other contributors to the perception of orientation are the vestibular system (see 1a.8) and the somatosensory system (pressure and position nerve receptors distributed throughout the body that provide information, for example, on the orientation of the seat in which we sit). There are two important classes of vestibular illusion that can produce inaccurate models of orientation (ie spatial disorientation).

The first concerns the perception of linear acceleration. The only force experienced by a pilot in level flight at constant speed is his own weight. If the pilot accelerates forwards an inertial force is produced by the acceleration that acts as shown in Figure 2b.2. This figure also shows that the resultant of these two forces in an accelerating aircraft operates similarly to the weight force in a climbing aircraft. Because the otoliths act very much as a spirit level, a pilot may readily mistake acceleration for pitch, a situation known as the somatogravic illusion.

If the pilot suffering from this illusion attempts to combat his perception of pitch-up by pushing forwards on the control column, the situation is aggravated since the weight force in the bunting aircraft is reduced and the resultant of the

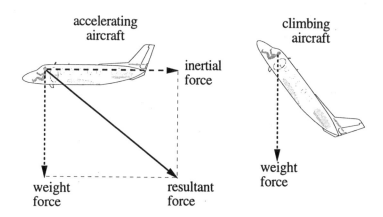

Figure 2b.2. The somatogravic illusion

weight and inertial forces rotates further aft – possibly producing a greater sensation of increasing pitch. The pilot may well attempt to counter this feeling by pushing further on the stick, increasing the bunt, and increasing the perception of pitch up – an extremely dangerous positive feedback loop. This pattern of events has undoubtedly accounted for a number of accidents, typically associated with acceleration during the overshoot from abandonned approaches in poor visibility.

The second class of vestibular illusion is concerned with roll. If a pilot enters a co-ordinated turn very gently, he may do so at a rate of roll that is below the threshold of detection of the semi-circular canals (which act as angular accelerometers). Since the resultant of the weight and turning force vectors is aligned with the pilot's head-foot axis in a co-ordinated turn, he may perceive his orientation as wings level. If he consciously appreciates this situation and rolls out of the turn at a rate capable of stimulating the semi-circular canals, he may fly back to a genuine wings level attitude, but feel as though he has banked away from wings level. In this illusion, known commonly as the 'leans', the pilot may have a conscious knowledge of his genuine orientation from the instruments or external world, yet retain a very compelling false feeling of leaning for a considerable time, durations of over an hour having been reported.

The only solution to all forms of spatial disorientation is for the pilot to trust the most reliable form of information available to him, and this will almost invariably be his instruments rather than his own sensations. This is the essential skill in all instrument flying and represents the only advice which may be given to pilots on this topic.

2b.3 Visual Cues in Landing

Probably the most critical visual tasks that pilots are presented with are the judgements involved in landing. These may be divided into three phases: initial judgement of an appropriate (eg 3°) glide slope, maintenance of the glide slope during the approach, and ground proximity judgements before touchdown.

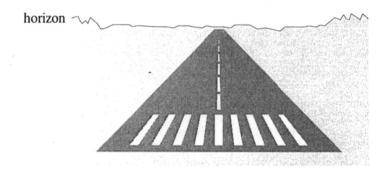

Figure 2b.3. The 'pilot's-eye view' of a runway

Initial judgement of an appropriate glide slope may be facilitated by aids such as visual approach slope indicators (VASIs), precision approach path indicators (PAPIs), or by positioning the aircraft at predetermined heights above known ground features (especially when circuit flying), but sometimes judgements without such aids must be made. When doing so the pilot is presented with an image on his retina that approximates to that shown in Figure 2b.3. If he wishes to fly a 3° approach then he must ensure that the angle, at his eye (the visual angle), between the impact point that he chooses on the runway (ie the point at which he visually aims) is 3° below the horizon. Figure 2b.4 illustrates why this is so. Since the line joining the pilot's eye to the horizon stays parallel with the surface of the earth (assuming an infinite flat earth), the visual angle between impact point and horizon is both constant and equal to the angle of approach.

It is often true that a pilot making such a visual judgement is prevented from seeing the horizon by poor visibility or because it is night. He may estimate its position in such circumstances by a number of means. The extended sides of the runway intersect at the horizon, and texture gradients in the surrounding terrain also indicate horizon location, but these cues are accurate only when terrain and

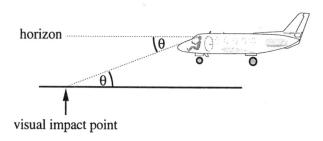

Figure 2b.4. Relationship between visual angle, visual impact point, and the horizon

runway are level. Sloping runways or terrain may produce incorrect estimates of horizon location in the pilot, and inaccurate approach slope judgements may result.

During visual glide slope maintenance, the pilot is required to aim for the visual impact point, and prevent the approach angle from varying. The pilot is enabled to know that he is aiming at his projected impact point, because this is the point on the ground away from which visual texture flows on the retina (see Figure 2b.5). As long as the visual texture flows away from this point, and as long as the visual angle between this point and the horizon remains constant the approach is progressing normally.

In the final phase, immediately before touchdown, the pilot may well be required to make height judgements above the ground to ensure that the

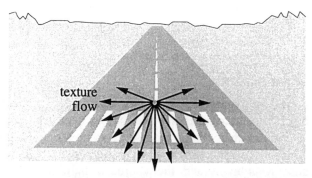

Figure 2b.5. Visual texture flow moving away from the visual impact point

undercarriage does not touch down before the runway threshold. Figure 2b.6 shows that the undercarriage will touch down a considerable distance before the visual impact point if the approach is continued without any check or flare. In order to make such height judgements, the pilot may use a number of cues that are likely to include the apparent speed at which ground texture is passing the aircraft (increasing apparent speed is produced by proximity), and the apparent size of ground texture and known objects on the ground (texture and objects grow apparently larger as they grow nearer). Both such cues are largely removed if the approach is over water, over snow, over other such featureless terrain, or carried out at night. In such circumstances, the pilot may be forced to make judgements based, for example, only on the apparent size (or width) of the runway. If the runway is narrower than he assumes it to be, he may interpret this reduced size as increased range, and thus touch down before he expects to. An important function of good approach lighting is to provide the pilot with cues in the undershoot that enable him to judge ground proximity.

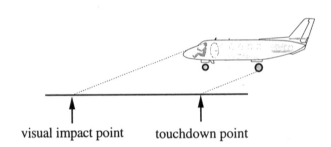

visual impact point   touchdown point

Figure 2b.6. Relationship between visual impact point and touchdown point

Consideration of the foregoing indicates that a particularly hazardous combination of circumstances is that which has induced a pilot to make a low (2° instead of 3°) approach (perhaps because of an up sloping runway), and that has prevented him from appreciating his ground proximity before touchdown (perhaps because it is night and the approach lighting is poor). The low approach means that the distance between the aiming point and touchdown point will have increased (see Figure 2b.6), and the failure to appreciate ground proximity means that the gear may impact the ground well before the threshold.

2b.4 Visual Search and Mid-Air Collisions

If two aircraft are to collide, then as long as they are unaccelerated (they do not change speed or heading) they will remain at constant relative bearings to one another in both azimuth and elevation (Figure 2b.7). This is true no matter what the headings or speeds of the aircraft, and is even true if they are climbing or descending. The practical consequence of this is that all aircraft with which the pilot is *not* going to collide will track across his windshield, and generate a movement cue that may assist detection, but the extremely rare aircraft that is on a collision course will remain stationary on the windshield, and will possess no relative movement to assist its detection. The situation is worsened if the aircraft presenting the hazard is located, from the pilot's point of view, behind an opaque windshield support since it will stay located behind such a visual obstruction until impact. This circumstance may well have prevailed in the DC9-Trident mid-air collision that occurred over Zagreb (1976).

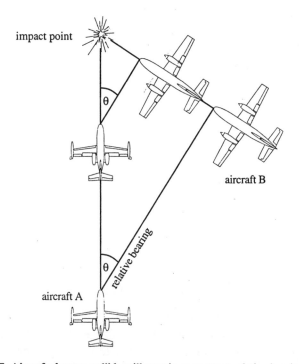

Figure 2b.7. Aircraft about to collide will remain at constant relative bearings
(ie there will be no movement across the windshield)

As the aircraft presenting the hazard closes, its retinal size increases. The rate of such increase has been plotted in Figure 2b.8 for a hypothetical aircraft approaching at a closing speed of about 800 knots. It can be seen from this figure that the image of the aircraft remains very small until very shortly before impact. The pilot maintaining lookout for other aircraft is thus presented with the problem of detecting a small and stationary target. The way in which he carries out a visual search for such a target is constrained by two properties of the eye. The first is that visual acuity (the capacity to resolve fine detail) is not distributed equally across the retina (see Figure 1a.6). Only the central part of the retina (around the fovea) possesses 6/6 or 20/20 vision, and that acuity drops rapidly even at small retinal distances away from this central area. In order to detect a small oncoming aircraft, the eye must be moved over the external world in order to provide the best chance of gaining this aircraft as a target on the foveal area.

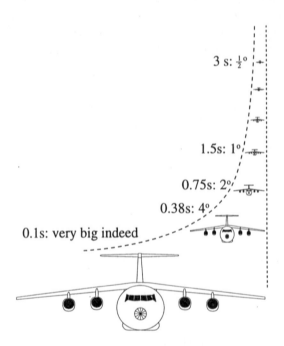

Figure 2b.8. The retinal size of an approaching aircraft before impact

The eye cannot be moved continuously and smoothly when it is searching for a target, but moves in jerks (saccades) with rests between them. The external world is sampled only during the rests, though eye and brain integrate the information acquired in order to provide a smooth perception of events. An eye movement/rest cycle takes about 1/3 second, and it can readily be seen that this means that the amount of external world that can be examined in detail is strictly limited. The pilot can maximize the probability of detecting a hazardous aircraft by minimizing the duration of the rests and making as many eye movements as possible. Although it is plainly important for the pilot to maintain lookout, it is probably true that however conscientiously he does so, detection of all colliding aircraft cannot be guaranteed. In a study that took place after a mid-air collision between a DC9 and a PA28 over Cerritos, California (1986), pilots were asked, during actual flight, to maintain a lookout, and another aircraft was deliberately flown towards them on a near collision course. Only 51 per cent of the experimental pilots saw the aircraft at all.

2b.5 Categories of Behaviour

The behaviour of any skilled operator such as a pilot may be broadly broken down into into three categories. Skill-based behaviours are those that rely on stored routines or motor programmes that have been learned with practice and which may be executed without conscious thought. Thus moving the control column, and operating the undercarriage lever may be regarded as skills, since the initial conscious intent is translated into actions automatically.

Rule-based behaviours are those for which a routine or procedure has been learned. The components of a rule-based behaviour (such as responding to a fire warning) may comprise a set of discrete skills (such as closing a throttle, shutting a start lever, and firing an extinguisher).

Knowledge-based behaviours are those for which no procedure has been established. These require the pilot to evaluate information, and then use his knowledge and experience (or airmanship) to formulate a plan for dealing with the situation.

2b.6 Skill-Based Behaviour

Skills may be acquired in distinctly different ways. The pilot may concentrate on individual aspects of the skill, giving conscious attention to these, until with practice these individual components are integrated into a smooth and automatic end product. Alternatively, the pilot may simply practise the whole pattern of behaviour, concentrating only on the accuracy of the final outcome, until eventually he lays down a 'motor programme' that produces the result he requires.

In fact, something of each of these approaches occurs in any process of skill acquisition. Once acquired, however, skills appear to possess certain characteristics. For example, the skill will normally be stored in the form of 'non-declarative' knowledge, ie the possessor of the skill may well not be able to articulate what the components of the skill are and this may cause difficulty if he wishes to pass the skill on to another person. What is more, if the owner of a skill wishes to modify it, he may well find that thinking about what he does spoils the execution of the skill, and he must go back almost to scratch in order to bring about change.

The decision to exercise a skill is normally made consciously by the central decision maker (Figure 2a.1), but the store of motor programmes appears then to be able to make use of sensory input and to have access to the neural mechanisms that control muscles without necessary reference to the central decision maker. This means that skills (such as walking, and some components of driving or flying) may be exercised at the same time as another activity that requires conscious control (such as maintaining a conversation with air traffic control). If, however, aircraft control becomes demanding and necessitates conscious decision making, then there will be no spare capacity in the central decision maker to maintain the conversation. In short, we are able to do two things at once so long as they do not compete for the resources of the central decision maker.

Ideally, any pilot exercising a skill, such as lowering the undercarriage would make the decision to do so, and then monitor his own behaviour in order to ensure that the correct skill was exercised. This may normally be so, but if the central decision maker is busy with another activity (ie the pilot is preoccupied), he may make the correct initial decision, inadvertently exercise the wrong skill, but fail to monitor his activity and remain completely unaware of the mistake that he has made. This mechanism of error is very common on flight decks, and examples abound of inadvertent control operations such as raising flaps instead of undercarriage immediately after take-off, shutting HP cocks on finals instead of lowering flap, or lowering flap instead of undercarriage during the approach. Many accidents have been caused by this mechanism.

A second route to error with a skill is that if it is frequently operated in the same environment (and becomes a habit), it may be elicited by that environment even though the pilot has not made a conscious decision to operate the skill (an effect known as 'environmental capture'). A good example of this is associated with the call 'Three Greens' made on finals. Simply being on finals may be enough to make the pilot make this call even though he has given no conscious attention to doing so and even though three green lights may not be present. Virtually all pilots who have landed gear-up have combined the errors described here, and have initially made a selection of, eg flaps, when they intended to select gear, and then called 'Three Greens' because this skill has become environmentally captured.

The pilot should be aware that errors of this sort are more likely when he is preoccupied (perhaps with a problem not associated with his flying), when he is tired (and central processing capacity is reduced), or when good conditions may have led him to relax too much. The pilot may be advised to endeavour to make all of his actions (and certainly vital actions) considered and deliberate rather than fast and slick, and possibly to discipline himself into carrying out actions (skills) in two stages. First, initiate the action and grasp the relevant control, and second, stop and consciously check that the correct control is about to be operated before continuing.

Lastly, it should be remembered that these errors of skill do not happen to novices since they have to think about what they are doing. They occur only to those with experience.

2b.7 Rule-Based Behaviour

It is probably true to say that it is the development of rule-based behaviours or procedures in aviation, and the associated training and checking of these behaviours, that has made commercial aviation as safe as it is. Every set of events that can be anticipated has been considered and reduced to a set of procedures or rule-based behaviours for the pilot to follow. Thus for example, the procedure to be adopted in order to land at Pisa has been considered and detailed in the Terminal Approach Plate so that the pilot does not have to work out from scratch how to do it. Similarly, the procedure to be adopted in the event of engine failure at V1 has been worked out and the procedure will be practised regularly by the pilot during simulation.

Rule-based behaviours are not stored as patterns of motor activity, but (as the name implies) as sets of rules, and are thus stored in our long-term memory. When they are actioned, however, they clearly require the involvement of our central decision maker and working memory as well since rule-based behaviours are always actioned at a conscious level. Working memory is clearly involved in carrying out a memorized procedure since we must keep track of what we are doing as the sequence of actions progresses.

It is in the area of procedural training that simulation is traditionally strongest and has been used most thoroughly. This is because many of the emergency procedures with which the pilot must be familiar cannot be trained in the aircraft, rarely occur in the aircraft, yet the pilot must remain in practice with them (ie his long-term memory must be refreshed and access to it exercised). Some procedures may be too involved to be reliably memorized, and thus are committed to some form of documentation such as flight reference cards. Even in these cases the pilot must retain a basic memory of the procedure that is required in order to access the correct information, and this must be practised. Generally speaking, procedures should be committed to documentation unless they will need to be exercised under circumstances that might prohibit, perhaps because of restricted time, document consultation. There is no virtue in memorizing rules unnecessarily.

Although these considerations indicate that rule-based behaviours are open to the sorts of failures that occur in long-term memory, in the central decision maker, and in working memory, they appear, in practice, to be very robust and to have many strengths. For example, standardized procedures enable each pilot in a crew to know what the other pilot should do in a given set of circumstances. The errors that occur in procedural behaviour are usually because the pilot has made an initial misidentification of a problem and engages the wrong procedure entirely. For example, he may have misidentified an auditory warning and thus engage the procedure for cabin depressurization instead of overspeed. Errors may also occur if the pilot believes that it is safe to depart from the procedure. For example, although it is always advisable, and in many airlines mandatory, to respond to the ground proximity warning with full power and pitch up, many aircraft have flown into terrain with this simple procedure having been ignored. Once a set of circumstances has been accurately identified, it is almost invariably advisable to exercise the standard procedure.

2b.8 Knowledge-Based Behaviour

Knowledge-based behaviour may, perhaps, be better termed decision making, or even thinking and reasoning. Decision making is clearly carried out by the central decision maker in Figure 2a.1, but equally clearly requires all of the information available to the pilot from his environment and memory. The fact that the pilot is able to evaluate evidence and come to conclusions will, in future, be the only reason for keeping him on the flight deck. The skill element of flying has been increasingly automated since the first autopilots, and the procedural element of flying can be very fully automated with modern computerized management systems, but the pilot must remain for the foreseeable future in order to think, reason, and evaluate the unexpected.

A very simple example serves to illustrate some important aspects of how people tend to evaluate evidence and make consequent decisions. Imagine you are told that there is a rule that connects together sets of three numbers, and that you have the task of discovering the rule. To help you, you are given an example of a set of three numbers that fits the rule; you can try out further sets of numbers and will be told whether they fit the rule or not. The example set of numbers is 2, 4, 6. A person playing this game will often try a set of numbers such as 10, 12, 14 and be told that the set fits the rule. He will try several similar sets of numbers and come rapidly (perhaps after only one trial) to the conclusion that the rule is n, n+2, n+4, only to be told that this is not the rule. Despite being told this, the future examples that the person tries are likely to be further instances that fit his own inference (eg 24, 26, 28), and when he is told that they fit the rule, he becomes more and more sure that n, n+2, n+4 is the right rule, and may even refuse to believe that it is not the rule.

Another person may come by a similar path, and equally rapidly, to the conclusion that the rule is n, 2n, 3n, try out such instances (eg 50, 100, 150), be told that they fit the rule, and be equally disappointed to be told that it is not the rule. The explanation is that the rule is 'Any three numbers in ascending order'.

A number of lessons may be learned from this game:-

a) Data may be ambiguous.

b) People are very keen to structure information and to make inferences from it.

c) The inferences that people make are very heavily influenced by their experience and by the probability structure of the data.

d) Once a person has formulated a certain way of thinking about a problem, it appears difficult for him to get out of that way of thinking and try a different interpretation of the data.

e) Even if a person tries to test his hypothesis about a set of information he is likely to try only positive instances of his hypothesis and unlikely to try negative instances of his hypothesis.

f) A point not apparent from the above example is that even if the person is presented with negative instances of his hypothesis, he is likely actually to disregard them (a process known as 'confirmation bias').

g) The final point (also not apparent from the above) is that people make inferences in accord with their wishes, hopes, and desires.

All of these points have importance in aviation and are now addressed in turn in this context.

a) The ambiguity of information presented on the flight deck has been the starting point of many accidents. For example, when the engine of the Boeing 737 at Manchester exploded (1985), the noise was heard on the flight deck, but could just as well have been a tyre bursting, and this was the interpretation made by the crew.

b) People seem to be unhappy to deal with information in an unstructured way and like imposing form. This is especially true on flight decks where having a hypothesis is most anxiety reducing by comparison with admitting to oneself that one does not understand what is going on.

c) Experience and judged probabilities are extremely important. In the Manchester accident, the most likely cause of a bang during the roll would have been a tyre burst, and this was thus the assumption of the crew until overwhelming evidence (a fire warning) was presented to the contrary.

d) Once a hypothesis has been made, people are usually happy with it until, as in the example above, evidence to the contrary becomes overwhelming. In a DC10 bound for Miami from Frankfurt (1979), the crew managed to engage the autopilot and autothrottle wrongly so that the aircraft slowed and approached the stall. The airframe buffet was interpreted by the crew as engine vibration (ambiguous data), they concentrated their attention on the engines and believed they had a problem with the No.3 engine. When the stick shaker indicated the impending stall, the crew interpreted even this as a rough running engine, and therefore shut it down. Only after the aircraft stalled (overwhelming evidence) did the crew come to appreciate the situation accurately.

e) The failure to try negative instances of the hypothesis is exemplified in the Kegworth 737 accident (1989). As soon as the first officer decided that the No.2 engine was causing problems, its throttle was closed. This was done at the same time as the No.1 (failing) engine starting to run smoothly. The low subjective probability of such a conjunction of events would have served to reinforce the crew's hypothesis, and they never contemplated attempting a negative instance of their hypothesis - ie they did not open the throttles of the No.2 engine and close the throttles of the No.1 in order to test their ideas.

f) Confirmation bias or the tendency to accept hypothesis confirming information and disregard hypothesis 'disfirming' information is well exemplified by the Boeing 747 which nearly landed in the grassland some miles short of the runway at Nairobi (1975). In this aircraft, the crew had been cleared to descend to 'seven, five, zero, zero' by the controller, but had not heard the 'seven' and interpreted the five followed by some zeroes as a clearance to five thousand feet. During the descent a number of indications on the flight deck should have told the crew that there was something very wrong (eg glide slope pointers out of view in the up position), but the crew interpreted these as aircraft failures and did not question their hypothesis until they had broken cloud and descended to very close proximity with the ground.

g) The effect of desire is shown by the accident to a Boeing 747 that flew into the tops of some rubber trees short of the runway at Kuala Lumpur (1983). An unusual descent had resulted in the rubber tree strike - a clearly undesirable outcome from the point of view of the pilots. The noises associated with the impact were interpreted as an engine surge by the pilots (a much more desirable possibility for them) even though the engineer believed that a ground impact had occurred.

2b.9 Situational Awareness

All of the foregoing has been concerned with ensuring that the pilot maintains accurate mental models of his environment, and this process is sometimes referred to as maintaining situational awareness. In order to ensure that the pilot maintains the best possible situational awareness, there are some guidelines that may be drawn from the above.

Gather as much data as possible from every possible source before making an inference (making up your mind).

Take as much time as is available to make your mind up (don't leap to conclusions), remembering that rapid decisions are seldom necessary.

Consider all of the possible interpretations of the data that you can think of - including the unlikely ones - before deciding which fits the data best.

Once you have embarked on a course of action try to stop occasionally to take stock of the situation.

Question whether your hypothesis still fits the data as events progress.

Consider ways in which you can test your actions to see whether your hypothesis is accurate.

If incoming data do not appear to fit your hypothesis, do not assume they are wrong or disregard them, but make the time to reconsider the situation, retracing your steps to the first signs of the problem if necessary.

Ensure that you do not interpret the world in terms of how you would like it to be, but in terms of how it is. By all means hope for the best, but plan for the worst.

PART III

STRESS, FATIGUE, AND THEIR MANAGEMENT

3a STRESS AND STRESS MANAGEMENT

Introduction

This section deals with the factors in the individual's work conditions and domestic situation that cause stress, models of its effects on the individual, and techniques for coping.

3a.1 Definitions, Concepts, and Models

When any physical object is acted upon by conflicting forces it is subjected to internal stress, and using the word stress to describe the internal state that is brought about in people by the pressures that life brings to bear on them is in direct analogy to this. The amount of stress which we experience influences both the way we feel and also our ability to perform tasks, but these effects can be both positive and negative. We can all recall instances when a small amount of stimulation or stress made us feel good and kept us on our toes, but, equally, we are also aware that under high levels of stress we can experience anxiety, depression, and other unwelcome effects.

The model of stress and coping in Figure 3a.1 illustrates the processes which cause an individual to experience stress. One of the features of stress is that an event which causes high stress in one individual may not have the same effect on another, and something which is stressful for an individual on one occasion may not be stressful on another occasion.

In any situation stress arises as a result of the evaluation individuals make of the demands which they believe to be placed upon them and the ability they feel they have to cope with these demands. Since the process of perception enables

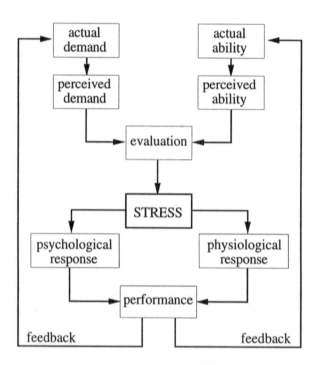

Figure 3a.1. A model of stress and coping

individuals to interpret and attach their own meaning to the information which they receive (2a.3), it is the individual's interpretation of the demands imposed rather than the actual demands which will be used in his evaluation. Equally, it is an individual's perception of his abilities rather than his actual abilities that contribute to the stress experience. We are all familiar with people who have far greater ability than they believe or are prepared to accept and who frequently show signs of stress when confronted with a task which is well within their capacity. Conversely, when an individual perceives a demand that he believes is well within his capabilities then the associated stress will be low.

High stress is associated with unpleasant psychological and physiological responses (eg fear, anxiety, sweating, fatigue). If the evaluation process causes an individual to decide not to undertake a difficult task, he will have to cope with the stress associated with refusing to meet the demand, for example, refusing to undertake a flight in an aircraft of doubtful serviceability or in marginal weather. Alternatively an individual may attempt to change a demand in such a way that he will feel able to cope. For instance, in bad weather, the pilot may be able to renegotiate the time at which the flight is to be made. If a demand cannot be changed and must be met, such as an emergency on the flight deck, the pilot will have to attempt to meet the demand even if high stress is experienced as a result of his evaluations of both the situation and his ability to control the aircraft. If a person successfully completes a task which he had initially perceived as being extremely difficult, his perception of his abilities to perform tasks of this nature will change. When faced with a similar demand in the future, he will be able to tackle it with greater confidence and a lower level of stress. Thus the stress associated with our perceptions of particular demands will continuously change as we improve or change our abilities to cope with the situation.

3a.2 Effects of Stress on Attention, Motivation and Performance

The way in which stress affects our performance of tasks is often conceived in terms of the intermediate factor of 'arousal' (3a.3). Many stresses are regarded as acting to increase our arousal level, and some as reducing it. The effect of arousal on performance is described simply in the model shown in Figure 3a.2. From the model it can be seen that, in the low arousal area of the curve, increasing arousal improves performance. At low levels of arousal, when we are not expecting to have to perform any difficult tasks, or when our motivation to perform tasks is low, our attentional mechanisms will not be particularly active. In this state we scan the environment slowly without focusing on any particular source of information and as a consequence occasionally miss information. When in this state, the fact that our performance capability is low may be

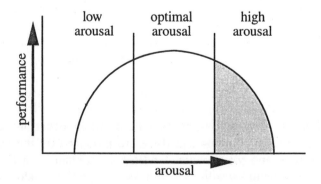

Figure 3a.2. The relationship between arousal and performance

irrelevant, for example, if we are relaxing at home at the weekend. If, however, our motivation to perform the task is low, perhaps because we are in a state of boredom and feeling sleepy in the cruise phase of a long-haul flight, the fact that we may miss incoming information because we do not scan the environment effectively can have serious consequences.

As the level of arousal is increased the ability to perform tasks reliably and accurately is also changed. The precise levels of stress which enable an individual to perform optimally will be a function of both task complexity and his experience and familiarity with the task. Although, generally speaking, complex tasks requiring a calm approach are performed best at lower levels of arousal, and simple tasks that can be performed energetically at higher levels of arousal, once the level of arousal becomes high, performance starts to deteriorate and people make errors. The deterioration in performance occurs, at least to some extent, because as arousal increases, we tend to restrict or focus our attention onto the event which we perceive to be the primary demand (an effect often termed 'narrowing of attention'). Thus errors can be made because information from possibly important, but more peripheral and non-attended, sources is missed (see 2a.9).

3a.3 Arousal, Concepts of Over- and Under-Arousal

As already outlined above, arousal is the mechanism used to describe the physiological responses made by the body to stress. Arousal can be thought of as a continuum of wakefulness ranging from very deep sleep to high levels of excitement. At low levels of arousal (under-arousal), such as shortly after awakening or during extreme fatigue, the nervous system is not fully functioning and the processing of sensory information is slow. Moderate levels of arousal produce interest in external events and performing tasks.

The arousal mechanism operates through the activity of the part of the nervous system known as the autonomic nervous system. The autonomic nervous system controls many of the functions essential to life, such as respiration and circulation, over which we normally have no conscious control, and is divided into the sympathetic and parasympathetic nervous systems. The sympathetic nervous system provides us with the resources to cope with a new sudden source of stress. This is known as the 'fight or flight' response and the purpose is to prepare the individual for physical activity. We are all familiar with the sudden increase in heart and respiration rates which occur in response to a sudden and unexpected stimulus and these are produced by the activity of the sympathetic system and the associated release of adrenalin. In addition, blood flow to the muscles and sweating are increased, digestion is slowed, and sugar from the liver is released for use.

Although these reactions were historically useful for survival when man was a hunter, they are usually inappropriate as a response to modern day sources of stress. One of the difficulties is that although our bodies are physically prepared for activity when stress occurs, we are frequently in a situation where we cannot release this physical charge of energy. Even when the source of the stressful experience has passed, our 'fight or flight' reactions to the source of the stress may continue. Furthermore the stress reaction can occur without the actual occurrence of the event but as a response to the anticipation of the perceived demand or threat.

3a.4 Environmental Stressors and their Effects

Repeated exposure to moderate levels of stress from the environment cause the body to adapt to the stress in order to reduce its impact (eg when we live next to a main road we cease to notice some of the traffic). On the flight deck, environmental stressors include excessive heat, noise, vibration, and low humidity. Individuals clearly differ in their tolerance to such environmental stressors, although the negative consequences of one source of stress are likely to lower an individual's resistance to other sources of stress. This point is illustrated by a CHIRP report from a helicopter pilot flying over the North Sea.

> After completing a round trip of 5hrs 05mins completely without incident, both my co-pilot and I felt absolutely knackered: we both had flown approx 150 hrs on similar routes in the past two months, which is not excessive by any standard currently likely to be recognized. However the excessive noise in the cockpit, the appalling vibrations and the noisy radio/intercom combined with the wearing of a survival suit combine to produce an accident waiting to happen. I believe that the stresses of long distance flying in helicopters are not recognized. These stresses are mental, not physical, and are by nature cumulative.

Heat

The comfortable temperature for most people in normal clothing is around 20°C. Above 30°C or below 15°C people begin to experience subjective discomfort, followed by physical discomfort and decreased efficiency. Above 30°C heart rate, blood pressure and sweat rate increase and attention is restricted or focused. Pilots would not normally expect to be exposed to temperatures above 30°C, although studies of helicopter cockpits have shown temperatures as high as 50°C. Low temperatures are also uncomfortable, and even if not apparently severe (eg in the case of a pilot ditched in the sea at temperatures as high as +10°C) can rapidly lead to incapacitation as a result of loss of feeling and control in the hands.

Noise

In conditions of low arousal - for example, when a person is suffering from lack of sleep - noise may actually improve performance. Noise may be used to help to maintain arousal levels during periods of boredom and fatigue. It may also be used to mask otherwise distracting sounds. However, excessive noise can disrupt the performance of a task, cause annoyance and irritability, and lead to cardiovascular and other physiological reactions. The sounds used for warnings on aircraft can produce just such responses, but recent research has aimed to design sounds that attract attention without causing a startle response from the pilot. Excessive noise has a similar effect to excessive heat in that attention tends to become focused and restricted.

Vibration

There is no doubt that working in vibrating environments for any length of time affects both visual and psychomotor performance. Physical symptoms are determined by both the frequency and amplititude of the vibration, but can include: 1–4Hz interference with breathing, 4–10Hz chest and abdominal pains, 8–12Hz backache, 10–20Hz headaches, eyestrain, pains in the throat, disturbances of speech and muscular tensions. Every effort should be made either to provide the pilot with a vibration-free environment or to isolate him as much as possible from the vibration with a properly designed seat.

Low Humidity

The air temperature outside a commercial jet aircraft is often in the range of -40 to -45°C. Since the air conditioning system draws in this air which contains almost no water vapour, the relative humidity within the aircraft can fall to around five per cent or even lower, whereas levels of relative humidity should be between 40 and 60 per cent for optimal comfort at normal temperatures. At these levels of humidity, there are liable to be physical reactions such as a drying of the mucous membranes of the eyes, nose, and throat. The problem is worst on long–haul flights on which there may be other effects associated with the manufacture of less urine and water retention. Because of these effects, the traditional advice to drink large volumes of fluid in low humidity environments may not be appropriate, and the pilot may be better advised to drink only that amount of fluid that he needs for comfort (avoiding large quantities of caffeine), and dealing with skin or lips with moisturising agents or water sprays.

3a.5 Domestic Stress and Home Relationships, Bereavement, Financial, and Time Commitments

Since the effects of stress are not confined to the event or experience which orginally caused the stress, stress at home can affect a pilot at work and equally, stress at work can affect a pilot's home life.

Domestic Stress and Home Relationships

Any pilot suffering from domestic stress should be aware that this can potentially affect their concentration and performance when flying an aircraft. This is illustrated in the following excerpt from a CHIRP report.

> For at least two to three years prior to the incident there had been a steady deterioration in the state of my marriage to the extent that I would get up in the morning unnecessarily early to get out of the house before my wife and child woke up. On this particular morning this did not occur and I was subjected to a non-violent but angry argument which left me emotionally boiling, a state I remained in throughout my drive to the airport, through flight planning and indeed up to the incident itself.

The pilot went on to describe how he made an error of skill (2b.6) because he was pre-occupied with his domestic problem.

There is good evidence for a relationship between stress and health (3a.7) and some for a relationship between accident involvement and domestic stress. Any change in a person's domestic situation such as divorce, marital separation and even marriage will be a source of stress. Additionally, frequent changes of partner or poor relationships with one's partner or other members of the family will affect an individual's overall stress level. A pregnancy, a son or daughter leaving home, a partner leaving or returning to employment can all lead to increased levels of stress.

Bereavement

The loss of a spouse or partner has been found to lead to higher levels of stress than any other experience. This is closely followed by the death of a close family member, personal injury and illness, and a major change in health of a family member.

Stress from all of these sources will be likely to cause loss of concentration and performance with, in some instances, loss of physical condition. For instance, one miltary pilot performed his second gear up landing in a period of six weeks on the day he had taken his wife into hospital for the birth of their child which was stillborn.

Financial and Time Commitments

The extended periods away from home that can result from the requirement to generate income can be a source of stress. Unsympathetic company policies and practices can lead to pilots returning home in an exhausted state, unable to contribute fully to family life. A particular source of stress for long haul pilots can be finding that they are a long way away when problems occur at home that require their urgent or frequent attention.

3a.6 Work Stress, Relationships with Colleagues and Management

Work Stress

For pilots, stress can arise at work both from the immediate situation on the flight deck, and from the general operating environment.

On the flight deck high stress will obviously be experienced if a sudden emergency occurs especially if the pilot is unsure how to react, or if he feels he has not been adequately trained. This source of stress is kept to a minimum by the use of simulator training, and it is important to appreciate that such training has two distinct functions. The first is to provide the pilot with practice in executing the skills and procedures that he will need to deal with real emergencies in the air. The second is to reduce the stress generated by the real emergency and to prevent it reaching incapacitating proportions by exposing him to the same, or very similar situations, in the simulator. Other parts of flying are stressful, however, and the handling pilot – even when very experienced – almost invariably shows signs of increased sympathetic nervous system activity during the approach and landing. The increase in heart rate shown during this phase of flight is often used as an indicator of pilot workload, but it should be remembered that this heart rate increase is more associated with the stress of flying the approach than the amount of activity or thought involved.

In many occupations long term quantitative overload has been shown to be related to stress symptoms. This is a factor which must be considered when organizing the flying roster and other aspects of company policy. Although flight duty times have led to control over flying hours and rest periods, in the competitive world, airlines are compelled to make maximum use of their staff and equipment. This may lead to the development of flight schedules which become impossible to achieve because of the inevitable delays caused by overcrowded airports and airways. Such tight flying schedules mean that any instances of technical problems will lead to time pressures and associated stress among the crew and ground staff. On occasions this source of stress will lead to errors of omission and lack of care over checking procedures and, hence, affect flight safety.

Pressure and conflict in any sphere are potential sources of stress. In the aviation industry today the financial pressure on companies can cause pilots to work under conditions of considerable pressure. The following CHIRP report clearly illustrates the dangers to aviation safety caused by financial pressure and lack of commitment to sufficient rest.

> I had been working a busy roster and was feeling tired when the company rang and said would I go to Venice, pick up some passengers, and fly

through the Alps to save time and money. I had legally sufficient rest time and after lunch we left. As it was too late to fly through the Alps we had to refuel at Nice on the return. At Munich the weather required an ILS approach down to 200 feet. I can remember thinking, when the inner marker activated, that there must be something wrong with the ground equipment and queried Control. They replied that the ILS was serviceable, but I was too high. After over-shooting, shaking my head to increase my concentration, the second ILS was successful.

If a management consistently exerts pressure on its employees to operate in ways that are more consistent with the short-term economic health of the company than with safety and good practice, then the company is likely to develop symptoms of 'organizational stress'. These symptoms include poor industrial relations, antagonism at work, high labour turnover and absenteeism, and, most importantly for the aviation industry, a high accident rate.

Relationships with Colleagues and Management

Poor relationships with colleagues on and off the flight deck represent an obvious source of stress; communication problems and relationships on the flight deck are considered in Chapter 4a. The policies imposed by management not simply in terms of the flying roster but also regarding promotion, career development and other matters can all be potential sources of stress. In one extreme example, the stress caused by the owner and president of an airline instructing all pilots to fly below the minima in bad weather, to take off with higher weights than those permitted, and not to take sufficient reserve fuel, indirectly led to three serious accidents within the company in a short period of time.

3a.7 Life Stress and Health, Other Clinical Effects of Stress

The examples of work and domestic stress so far considered can all be regarded as part of general life stress. Life stress is sometimes referred to simply as 'life change' because it is widely believed that all life changes – even those that are apparently beneficial – require the change to be coped with and are thus stressful. The importance of a wide variety of life changes has been assessed, and a sort of league table can be drawn up with the most stressful at the top (death of child or spouse) scoring, say 100 points, and the least stressful (eg parking fine) at the bottom scoring 10 points. It has been found that the effects of these changes on the individual tend to be cumulative, to the extent that the more points you accumulate in a given period, the more likely you are to suffer from stress related illness. In one study, over half of those scoring more than 200 points in a two year period had their health affected.

In a study conducted on stress among pilots the main sources of stress identified were:

    a) lack of control or disruption of events in their lives

    b) scheduling and rostering

    c) anxiety about courses and checks

    d) home-to-work interface

    e) career and achievement

    f) insufficient flying

    g) lack of responsibility and decision making

    h) interpersonal problems

    i) management and organization issues

    j) domestic status

    k) fatigue and flying patterns.

Insufficient flying in the above list was clearly determined by the commercial environment prevailing at the time of the study: aviation seems always to have either too many pilots who are both flying too little and worrying about job security, or too few who are flying too much and worrying about the professional and domestic effects this is having on them.

Although the physiological and psychological bases of the effects of stress are highly complex and are still imperfectly understood, our actual reactions to stress are extremely basic and common. The human body appears to be programmed to mobilize defence actions against stimuli which disrupt our optimal physiological and psychological well-being. The possible effects of stress can be classified into five areas:

    Physiological effects - these include short term changes in response to a sudden unexpected event brought about by the action of the sympathetic nervous system (3a.3), such as dryness of the mouth, sweating, increased heart rate, and difficulty in breathing.

    Health effects - the symptoms most frequently associated with stress are those associated with the gastrointestinal system, eg nausea, indigestion, diarrhoea, and ulcers. The association between stress and coronary heart disease and permanently raised blood pressure is less clearly understood but there is evidence to support the idea that certain personality types (especially those who are competitive, who set difficult goals for themselves, and who are frustrated by failure) are at increased risk. Other health effects can include asthma, headaches, sleep disorders, sexual disorders, neuroses, allergies, colds and flu.

Behavioural effects - these can include the immediate responses of restlessness, nervous laughter, trembling, impulsive behaviour, excitability, and taking longer over tasks. Stress can also lead to excessive changes of appetite, and excessive drinking and smoking. An important effect of stress which all pilots should understand is that people suffering from stress are prone to accidents, thus particular care should therefore be taken by anyone suffering from stress when they are driving cars or flying aircraft.

Cognitive effects (thought processes) - the main effects of stress are lack of concentration, forgetfulness, inability to determine priorities or make decisions, and difficulty in switching off.

Subjective effects (feelings) - these can include anxiety, aggression, depression, fatigue, apathy, moodiness, tension, and irritability.

Although there are enormous differences between individuals in their responses to stressful situations, the extent to which stress from either domestic events or an individual's work situation gives rise to stress symptoms will be a function of the individual's personality, and the desirability and impact of the event. Not surprisingly, an event is likely to be perceived as stressful when it is perceived as a negative life change rather than a positive one.

3a.8 Coping Strategies, Identifying Stress, and Stress Management

We all possess a repertoire of psychological mechanisms or coping strategies which we use to combat stressful or unpleasant experiences.

Coping Strategies

Coping is the process whereby the individual either adjusts to the perceived demands of the situation or changes the situation itself. Some coping strategies appear to be carried out unconsciously: it is only if they are unsuccessful that we consciously take notice of a stressor.

Among the population a personality dimension or continuum of 'stress awareness' can be recognized. At one extreme there are individuals who repress knowledge of problems and thereby appear not to perceive them or require stress coping strategies: at the other extreme are those individuals who are extremely sensitive to problems to the extent that they will anticipate difficulties not perceived by others and employ coping strategies to avoid the experience of stress. Coping strategies can be classified into:

Action coping - Using this strategy the individual attempts to reduce the stress by taking some action, ie he reduces the level of demand in the model (Figure 3a.1). This may involve either removing the problem or altering the situation so that it becomes less demanding, or the individual may actually remove himself from the situation. Thus changing job, getting divorced, or refusing to comply with company pressure can all be regarded as examples of action coping.

Cognitive coping - Although action coping appears to be highly desirable, some situations cannot be changed. Cognitive coping involves reducing the emotional and physiological impact of stress on the individual. Some cognitive strategies that have been suggested include 'defence mechanisms' which operate outside conscious awareness, such as repression and denial. It is suggested that unconscious activity prevents the conscious brain from even becoming aware of the existence of a stressor. Other more obvious conscious strategies involve rationalization, or emotional and intellectual detachment from the situation. These processes change the perceived magnitude of the problem even if the demand itself has not changed.

Symptom-directed coping - This may involve the use of drugs such as tranquilizers, tobacco, alcohol, and even tea or coffee. In order to remove the symptoms of the stress other types of symptom-directed coping can include physical exercise, meditation and other stress management techniques.

Stress Management

Stress management is the process whereby individuals adopt systems to assist their coping strategies. The success of an individual's stress management will be determined by his willingness to recognize the sources of his stress and his determination to 'do something about it'. Some of the stress management techniques which can be useful include:

Health and fitness programmes - Regular physical exercise appears to reduce tension and anxiety. The physiological and biochemical changes which occur during exercise may also result in improved cognitive functioning.

Relaxation techniques - Meditation, self-hypnosis, biofeedback techniques, and 'autogenics' may all be regarded as forms of relaxation. Such methods frequently involve progressive muscle relaxation and mental imagery. Biofeedback techniques allow the individual to control his physiological relaxation by monitoring functions affected by sympathetic nervous system activity such as heart rate and blood pressure. The technique has been reported to help individuals to reduce anxiety, lower stomach acidity, and control tension.

Religious practices - There is no doubt that for many people some form of regular religious practice will help them to cope with stress especially following a major life event such as a bereavement.

Counselling techniques - Counselling may involve anything from attending sessions with a professional counsellor to simply talking about a stress problem with a supportive friend or colleague. The basic principle behind counselling is that since stress is caused by an individual's perceptions of a situation, the stress will be reduced if the individual can change the way he perceives or reacts to the situation by changing or modifying his beliefs or assumptions about events (cognitive coping). Counselling may also, however, assist an individual to see that some behavioural change (action coping) may be necessary, and help to bring about that change.

It should finally be noted that individuals, such as pilots, who must demonstrate authority and control in their work may show some initial reluctance to 'admit' that they are experiencing problems with stress since they may feel that it will be interpreted either as a lack of competence or weakness. Nevertheless, all pilots should be aware that stress can potentially influence performance on the flight deck - especially in an emergency - and take steps to deal with stress if they feel themselves to be affected by it.

3b SLEEP AND FATIGUE

Introduction

This chapter deals with issues relating to sleep and fatigue, and includes the basic concepts of biological rhythms, sleep/wakefulness cycles, and the nature and function of sleep. These areas are then applied to aviation with relation to time zone crossing, shift work, and sleep management.

3b.1 Biological Clocks and Chronobiology

Many physiological processes in the body exhibit regular rhythmic fluctuations, and they occur whether you are kept awake or allowed to sleep. These rhythms are controlled internally (endogenous rhythms) and are not simply reactions to the external environment. The most common rhythms exhibited by both man and other organisms have periodicities of or about 24 hours. These rhythms are termed 'circadian rhythms', from the Latin 'circa'- about, and 'dies'- day.

To make direct studies of endogenous rhythms scientists use isolation suites (bunkers) in which volunteers are given no external time cues. Typically, they are fed small meals at two or three hour intervals, and no clocks or watches are permitted so they cannot judge the time of day. In these conditions rhythms are said to 'free run'. Free running circadian rhythms normally exhibit a periodicity of 25 hours (like a slow running clock), so in isolation suites the free running time gradually moves away from true time – Figure 3b.1. In normal conditions our biological clock is locked to 24 hours by the onset of day and night, by clock times, and by social activities eg meal times. Cues such as these that serve to synchronize endogenous rhythms are known as 'zeitgebers' (German for 'time givers').

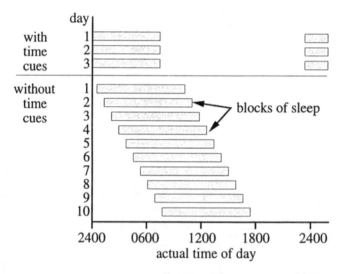

Figure 3b.1. Illustration of how an individual's body clock will run to 25 hours. When no time cues are present the subject's sleep period begins about an hour later each day.

The discipline of discovering and investigating biological rhythms is called chronobiology. As well as circadian rhythms we also exhibit rhythms of different periodicities, eg 28 day menstrual cycles and 90 minute sleep cycles (see 3b.3).

3b.2 Circadian Rhythms and the Sleep/Wake Cycle

The most commonly studied rhythms are those of activity (sleep/wake) and body temperature. Although body temperature is regarded as 37°C it does in fact 'normally' vary from 36.9°C in mid-evening to 36.2°C in the early morning (see Figure 3b.2). The temperature rhythm remains fairly constant in isolation studies; its periodicity stays close to 25 hours, and it cannot easily be entrained to different periods. For this reason the temperature rhythm is often used as a standard 'reference' rhythm to compare to others. By contrast the sleep/wake cycle can be entrained to between 20 and 28 hours. This provides evidence to suggest that there is more than one internal (body) clock.

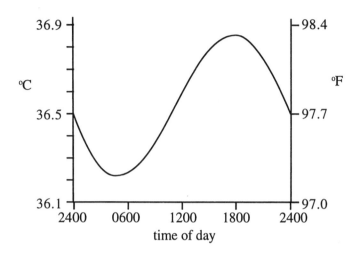

Figure 3b.2. The circadian rhythm of body temperature

In normal conditions the sleep/wake cycle follows a 24 hour rhythm with approximately one third of this time spent asleep. The body temperature rhythm and sleep/wake cycle run together such that the lowest point of the temperature rhythm coincides with the lowest point of the sleep/wake cycle (around 0500h). Generally, it is around this time that it is hardest to stay awake. It should be noted, however, that the temperature rhythm rises and falls as shown whether or not the individual takes any sleep.

Isolated from the body temperature rhythm the sleep/wake cycle can be thought of, very crudely, as a 'credit' and 'deficit' system in which a person is given two points for every hour asleep and has one point deducted for every hour awake. However long the sleep, the maximum of 16 points is never exceeded, ie you cannot 'store' sleep. The fewer points you have, the more ready you are for sleep. Normally a person will sleep when he has little or no sleep credit (ie around zero points), and will then sleep for about eight hours (16 points credit). This will be followed by a wakeful period of about 16 hours (16 points deducted), so at the end of a 24 hour period he will be back where he started and be ready for another period of sleep. This is illustrated in Figure 3b.3. From this it is easy to see how the credit system is affected by irregular work patterns, including extended periods of time awake with reduced or broken sleep between. A gradual reduction in the level of credit may build up over time. Such an accumulation of sleep loss is called a 'cumulative sleep debt'.

Although time awake is important in determining readiness for sleep, there is also a circadian rhythm of sleepiness. This means that at certain times of day even the sleep deprived individual may have difficulty in falling, and staying,

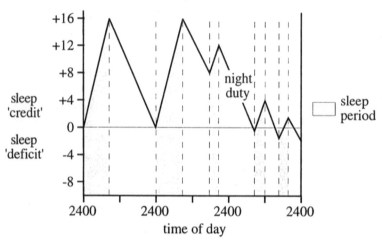

Figure 3b.3. The effect of a sleep/wake pattern on the credit/deficit system
(1 hour asleep = +2 points, 1 hour awake = -1 point)

asleep. It is the *timing* of sleep, not the amount of time awake, that is the *critical* factor influencing sleep duration. Because the duration of sleep is closely related to the phase of the body temperature rhythm, sleeps taken at times near the temperature peak (or when body temperature is falling) will be longer than sleeps taken at times near the temperature trough (or when body temperature is rising). This explains why sleeps taken at 'abnormal' times of day may be more disturbed and less refreshing, eg sleep attempted at 0700h will be less sustainable and invigorating than sleep attempted at 2200h.

A person's physiological sleepiness may be measured using a procedure called the Multiple Sleep Latency Test (MSLT). The person is put to bed in a darkened, quiet room and is instructed to sleep. The test stops after the first 90 seconds of sleep (confirmed by EEG), or 20 minutes, whichever is sooner. This procedure is repeated at regular intervals. The times taken to fall asleep are plotted and from this any trends in sleepiness can be observed.

Time of Day and Performance

As well as rhythms for basic physiological measures there are also circadian rhythms for more complex behaviours. Performance on different tasks is affected differently by time of day. Simple tasks that require low working memory loads (see 2a.5), eg simple reaction times, vigilance, visual search speeds, and manual dexterity, follow a pattern similar to that of body temperature, ie they improve during the morning and early afternoon, and decline during the evening to a low around 0400h. Performance on short term memory tasks, eg remembering phone numbers, declines through the day, and performance on mental arithmetic tasks and verbal reasoning tasks peaks around midday. However, as with many of these rhythms, there are marked differences from person to person.

The effects of time of day on accident rates have been studied extensively. Driving accidents have been found to peak at certain times of day, eg 1500h, and the effects of time of day have been noted as possible causal factors in a number of incidents and accidents, eg the aircraft incident in Nairobi (1975), the accident at Three Mile Island nuclear power station (1979), Chernobyl (1986), and the Challenger space shuttle disaster (1986). It is also interesting to note that in the BAC 1-11 incident where incorrect bolts were fitted in the captain's windshield (1990), the engineer fitted the windshield between 0300h and 0500h, probably during the lowest phase of his circadian efficiency.

3b.3 Sleep and Sleep Recording

When we are asleep we are in an altered state of consciousness, also described as a healthy state of inertia, or an unresponsiveness imposed on us by the nervous system. No one knows the precise function of sleep although much has been learned about the nature of sleep.

The discovery, 50 years ago, that brain activity could be recorded by attaching electrodes to the scalp enabled scientists to investigate the nature of sleep. In a sleep laboratory volenteers will typically have three types of measurement taken. Firstly their brain activity will be recorded using an electroencephalogram (EEG). The EEG comprises a number of electrodes that are glued to the head in particular positions. Secondly, eye movements are recorded using an electrooculogram (EOG) - electrodes being fixed to the outer corners of the eyes - and thirdly, electrical activity in the chin muscles is recorded using an electromyogram (EMG) - two electrodes are attached to the skin under the chin and this provides information about the muscle tension or relaxation. The electrical outputs from the EEG, EOG, and EMG are then recorded either on paper or into a computer.

Stages of Sleep

The stages of sleep are determined from the patterns of the EEG, EOG, and EMG activity. Sleep can be divided into five stages; Stages 1 to 4 and Rapid Eye Movement (REM) sleep.

When a person is awake the EEG shows two basic patterns of activity - alpha and beta. Alpha activity is observed when a person is resting quietly. This activity may be recorded from a person whose eyes are open but is more often recorded when the eyes are closed. Beta activity is seen while the person is alert or aroused and often takes over from alpha activity when a person is asked to solve a problem or if a loud noise is made.

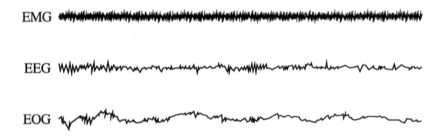

Figure 3b.4. Output from the EMG, EEG, & EOG of a subject in stage 1 sleep. (EOG – slow rolling eye movements, EEG – small, rapid irregular waves, EMG – active)

As the person starts to fall asleep alpha activity gives way to small, rapid irregular waves and the EOG shows slow rolling eye movements. This is Stage 1 sleep, a transitional phase between waking and sleeping (see Figure 3b.4). As sleep progresses the EEG contains increasing amounts of low frequency, high voltage activity (delta activity). Sleep stages 2–4 are largely defined by the amount of delta activity recorded, with the deeper stages of sleep (stages 3 and 4) having increasing amounts. Stages 3 and 4 are often referred to as slow wave sleep. The remaining stage of sleep is called rapid eye movement (REM) sleep, when this occurs the EEG becomes irregular (desynchronized), the EOG shows the eyes rapidly darting back and forth, and the EMG becomes silent indicating muscle relaxation (see Figure 3b.5). This sleep is sometimes termed 'paradoxical' sleep because of the EEG becoming similar to that of somebody who is awake.

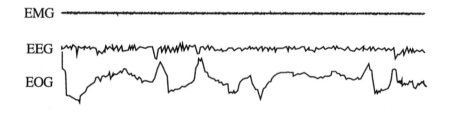

Figure 3b.5. Output from the EMG, EEG, & EOG of a subject during REM sleep. (EOG – darting eye movements, EEG – small, rapid, irregular waves, EMG – little activity)

Sleep Cycles

A person falling asleep will typically go down through the sleep stages spending around 10 minutes in stage 1 sleep, followed by 15 minutes in stage 2 sleep (about 50% of sleep is stage 2 sleep), and 15 minutes in stage 3 sleep, before moving on to stage 4. Approximately 90 minutes after sleep onset REM sleep will occur. The cycle of REM sleep and stage 1–4 sleep repeats during the course of the night in 90 minute cycles, each succeeding cycle containing greater amounts of REM sleep. An eight hour sleep will contain around four or five bouts of REM sleep. Most stage 4 sleep is accomplished early in the night. Figure 3b.6 shows a plan of the stages that a person experiences in a typical night's sleep.

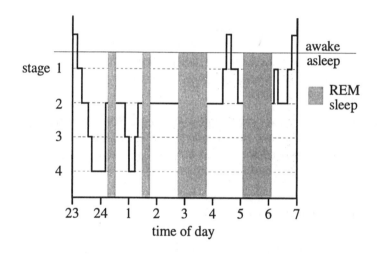

Figure 3b.6. Sleep profile for a typical night's sleep.

(Individual moves up and down through the sleep stages during the night. Four complete non-REM/REM cycles occur during the night.)

3b.4 Functions of Sleep

Although little is known about the precise function of sleep there are some differences identified in the function of slow wave and REM sleep. Neither of the two types of sleep should be regarded as deeper than the other – they are different. Awakenings from slow wave sleep often result in descriptions of situations or sensations, eg being crushed. Sleep walking and nightmares are more likely to occur early in the night during slow wave sleep. If awakened from REM sleep people will frequently remember more complex, bizarre, and emotionally-coloured dreams.

It has been suggested that slow wave sleep and REM sleep have different functions. Slow wave sleep may be related to body restoration, after you have carried out strenuous activity, eg intensive sports training, your body is likely to take more slow wave sleep. REM sleep may be related to strengthening and organizing memory - learning new tasks may result in an increased proportion of REM sleep.

Sleep deprivation experiments have shown that if a person is deprived of either slow wave sleep or REM sleep he will show rebound effects, ie he will make up the deficit in that particular type of sleep in subsequent sleep. This in turn implies that the body has some 'need' for these types of sleep.

Required Quantity of Sleep

Individuals require different quantities of sleep. In a study of one million people the most frequently reported sleep duration was between eight and nine hours. Shift workers generally average less sleep than non-shift workers, and UK charter pilots, who claim to need 7.5 hours sleep a night, report an average duration of 6.25 hours for sleeps obtained before duty days and 7.11 hours for sleeps obtained before rest days.

As you get older your sleep requirement changes. Small children take several periods of sleep in a 24 hour period whereas older children and adults tend to sleep in one continuous period. Infants have a high proportion of REM sleep, but this drops in adulthood and generally continues to decline into old age.

As people get older they sleep less, but with increasing age there is less flexibility about when this sleep is taken. This means that shift work becomes more difficult with age but these changes affect different individuals in different ways. Researchers have also looked for differences in the sleep requirements of men and women - although women generally have more sleep than men, they also report more sleep problems. In general, there is no absolute amount of sleep that must be achieved. You should sleep as much as you need.

3b.5 Shift Work and Sleep at Abnormal Times of Day

Sleep loss, or partial sleep deprivation is an occupational hazard of commercial flying and any other form of shift work. Inevitably there will be times at which you will have to work when you would rather be asleep, and conversely, times at which you have to sleep when you would rather be awake. It is at these times that sleep problems may be aggravated by circadian rhythms. As previously noted in 3b.2 the sleep/wake cycle affects your readiness for sleep and the timing of sleep, relative to your body temperature, is critical in influencing the duration of your sleep.

For example, if you are rostered for a night duty and you had a full night's sleep the previous night you will find it difficult to sleep well during the late afternoon before the duty. This is because you have a good sleep credit and an increasing body temperature both of which will work to make getting to sleep and sustaining sleep more difficult.

The effects of sleep disruption vary from person to person. In the situation described above, one person may find it easier to go to bed late the previous night, sleep in late in the morning before the duty and relax in the late afternoon so they still have a good sleep credit at the start of duty. Another person, however, may find it more beneficial to shorten his sleep the previous night and therefore have a smaller sleep credit in the afternoon increasing the likelihood that he will be able to sleep before his duty. The latter method has the drawback that if, for some reason, sleep is prohibited - eg the next door neighbour mows his lawn - there is the chance that the person may go on duty with an even greater sleep deficit.

The method you choose may depend on the type of person you are. The way various circadian rhythms interact gives rise to the phenomenon of morningness/eveningness. Morning people are early risers whose performance peaks early in the day, evening people are usually late risers whose performance peaks later in the day. Some people fit into neither category. Using the methods above the 'morning' person may benefit more by rising early whereas the 'evening' person may prefer to go to bed later and sleep in.

Another effect that shift work may have on sleep is suggested by the rebound effect in sleep deprivation experiments. From this it can be seen that an early call will shorten sleep such that REM sleep may be reduced by a large proportion, and this in turn may shorten the following night's slow wave sleep. There is no evidence, however, that this will have an adverse effect on the individual, and indeed, is a good example of the body's ability to regulate itself in a changing environment.

3b.6 Time Zone Crossing, Circadian Dysrhythmia, and Resynchronization

The problems of crossing time zones and jet-lag are a way of life for long-haul pilots. Time zone shifts can lead to considerable cumulative sleep deprivation in some adverse tours of duty. Although such sleep deficits can build up they are unlikely to get to extreme levels since the body will sleep when it needs to even in the least likely places.

Long-haul pilots have constantly to adjust and readjust their circadian rhythms. It is possible that continual disruption may incur some physical penalty and suggestions have been made of associations with stomach and bowel disorders.

The main problem of both time zone shifts and shift work, however, is that they cause desynchronization of body rhythms. For example, the normal rhythms of the alimentary canal and urinary system can serve as a source of sleep disruption in new time zones. The associated shifting of zeitgebers helps resynchronization to the new local time, but as most rhythms resynchronize at a rate of about one to one and a half hours a day (and some are slower) many pilots will not resynchronize in the new time zone unless they are stationed there for extended periods. There may be cases of some members of long–haul flight crews never attaining circadian stabilization throughout their flying careers, except during periods of leave or extended training in one place.

Differences in ease of resynchronization can be demonstrated using the example of flights to the west coast of the US. Most flights from UK to the west coast occur during the day. With an 1100h departure time from UK the pilot arrives at Los Angeles at 1400h US time (2200h UK time). The fact that the body temperature rhythm free runs to a period greater than 24 hours helps to enable the pilot to extend his day and stay awake for longer. This effect is combined with the effect of local zeitgebers, eg daylight, which are also working in the right direction to extend the day. The pilot therefore, will probably cope reasonably well with this time shift, and experience a good initial sleep. His second sleep period, however, may be poorer than his first, because he will have been awake for only a normal waking duration, will consequently not be in sleep debt, but will be trying to sleep at a time of low circadian sleepiness.

The situation for the pilot returning to UK is not so favourable. Flights back to UK often leave the US in order to arrive as UK airfields open. A typical flight might leave at 1400h US time and arrive 0700h UK time. After such an overnight flight the pilot will arrive sleep deprived, may therefore gain a reasonable initial sleep, experience a poor subsequent sleep because his circadian rhythm is out of phase with local day, and take a long time to resynchronize his circadian rhythms because they must be shortened in order to match his environment (ie against their natural tendency to run long).

3b.7 Rostering Problems, Sleep Management, and Naps

The sleep disruption caused by shift work and time zone crossing may be further complicated by rostering and sleep management difficulties. The pilot flying from the UK to the eastern or central US is likely to have a 24 hour slip at his destination before returning. If he arrives at 1500h US (perhaps 2100h UK) time, he will be ready for sleep when he arrives at his hotel. Should he take a full sleep at this time he will wake naturally about 15 or 16 hours before his return flight, and hence have a sleep credit of zero just as he is getting airborne. The only solution is to split his sleep into two shorter periods – the first after his arrival, and the second before his departure. If, however, he is unable to get to sleep during the second of these two periods, he may have been awake for twenty hours or so, and hence have a considerable sleep debt even before he takes off. All of these problems are exacerbated by attempting to sleep at times of day when hotel cleaning and daylight may make sleep difficult to accomplish.

Naps may have a beneficial effect for the pilot who has to work at an unusual time of day. Again some people find it easier to nap than others, with those people who nap habitually being more able to gain from a nap than non-nappers who may often suffer from 'hang over' effects (eg extended drowsiness, headaches etc).

3b.8 Sleep Disorders

The main disorders associated with sleep are narcolepsy, sleep apnoeas (or apneas), sleep walking and talking, and insomnia.

Individuals suffering from narcolepsy are individuals who are unable to prevent themselves from falling asleep, even when in sleep credit, and even in dangerous situations such as driving and flying. This is a recognized disorder and is clearly undesirable in aircrew.

Apnoea means cessation of breathing, and normal individuals may experience such apnoeas lasting around 10 seconds a few times a night. Sleep apnoeas tend to increase in frequency with age and may become a problem in some individuals (especially those with a history of heavy snoring) by becoming prolonged (lasting up to a minute), and excessively frequent (perhaps occurring hundreds of times a night). Frequent awakenings associated with this condition are likely to cause excessive daytime sleepiness, and since other clinical problems may be involved, medical advice should be sought.

Sleep walking (somnabulism) and sleep talking (somniloquism) are very common in childhood. They are still relatively common phenomena in adults and may occur more frequently in those operating irregular work/rest schedules. They should not cause difficulty in otherwise healthy adults, though (rarely) sleep walkers are involved in accidents.

Insomnia (difficulty in sleeping) may be divided into two categories – clinical and situational. Clinical insomnia (inability to sleep in normal, favourable conditions) may be overestimated by the sufferer since there is no absolute required quantity of sleep, and those who experience a 'sleepless night' have generally slept for considerably longer than they believe. Situational insomnia is the term used to describe inability to sleep due to disrupted work/rest patterns and is the problem frequently reported by aircrew. Some ways of dealing with both types of insomnia are described below.

3b.9 Sleep Hygiene

If your body really needs sleep it will sleep under almost any conditions. If, however, it is still in sleep credit, or if you are trying to sleep at a time of low circadian sleepiness, it may be useful to observe the following guidelines (sometimes collectively referred to as 'sleep hygiene'):

> Avoid drinks containing caffeine near bedtime
>
> Avoid napping during the day
>
> Make sure that the bed and room are comfortable
>
> Avoid excessive mental stimulation, emotional stress, or strenuous exercise before retiring
>
> A warm milky drink, light reading in bed, and simple progressive relaxation (see 3a) may well be helpful.

3b.10 Management of Sleep with Drugs

As individuals' tolerance to sleep disturbance differs some people will have to resort to the use of drugs to obtain sleep. The ability to gain sleep when the body is out of phase with local time is an important requirement for long–haul crews and for any pilot flying at night. Problems of sleep management can also occur in scheduled flying when there are incidents of roster disruption, staff sickness, and prolonged periods of 'standby', when crew members may suddenly be required to report for duty at a time when they are becoming ready for sleep. At times like this sleep may need to be delayed and the commonest 'drug' used for this purpose is caffeine.

Drugs sometimes prescribed for use by aircrew to promote sleep (called hypnotics) are members of a common group of drugs called benzodiazepines. These are best known under their trade names of Valium, Librium, Dalmane, Normison, and Mogadon, but some of these drugs may have serious effects on performance. An important characteristic of the drug is its half–life (the time it takes the drug to fall to one half of its peak level in the blood), since this determines the possible accumulation of the drug with continued use. Needless to say, it is of extreme importance that if you request sleeping drugs for use in an aviation context, the GP is aware of the fact that these drugs will be used in controlling situational insomnia (inability to sleep due to circadian de-synchronization) rather than clinical insomnia (the inability to sleep in normal, favourable conditions). The drug generally accepted as being the most suitable for pilots is, currently, Normison.

Finally, it should be noted that although alcohol is widely used by aircrew as an aid to sleep, it is a non-selective central nervous system depressant, and is plainly a drug. It may induce sleep, but this sleep is not normal since REM sleep is reduced considerably and early waking is likely. Other drugs not designed for use in sleep control, eg cold remedies and antihistamines, may cause drowsiness as a side effect, but the use of these drugs to induce sleep is strongly discouraged.

PART IV

THE SOCIAL PSYCHOLOGY

AND

ERGONOMICS OF THE

FLIGHT DECK

4a INDIVIDUAL DIFFERENCES, SOCIAL PSYCHOLOGY, AND FLIGHT DECK MANAGEMENT

Introduction

The following two sections address the ways in which pilots handle two interfaces – that between themselves and other people, and that between themselves and the equipment that they are operating.

4a.1 Personality and Intelligence

People differ from one another in many respects such as size, strength, hair colour, personal characteristics, and intelligence. Some of these individual differences are irrelevant in aviation, but many are not. Size and strength considerations must be taken into account in the design of controls and the cockpit workspace (see 4b), and others such as personality differences and intelligence are important in determining the role of the pilot, and the way in which pilots interact with one another.

Intelligence

The definition and measurement of intelligence is an area of unresolved contention, that fortunately need not cause difficulty for the practising pilot. Pilots are not generally selected for either training or employment in terms of raw intelligence tests but in term of tests of attainment, whether these be school examination passes or a commercial flying licence. Although intelligence may be required to attain such qualifications, it is the qualifications themselves that count.

A difficulty with this approach is that two pilots with the same qualifications may plainly differ in basic intelligence, and may therefore differ in the mental agility and reasoning power that they are able to bring to solving problems on the flight deck. Since there is little that can be done about this problem by the pilot, however, it is not pursued here.

Personality

The same may not be said of personality. The term is used to embrace all of those stable behavioural characteristics that are associated with an individual and which can be extremely important in determining his relationship with others even in the structured environment of the flight deck.

4a.2 Assessing Personality

It is commonly but accurately asserted that all of us are assessing personality in any social encouter that we have. There are over 17 000 words in English that are used to describe personality related behaviour, but some form of assessing personality rather more scientifically than only using everyday language is clearly desirable. There are a number of ways of attempting to do so. The first is by interview. Although personality judgments are made in interviews, these are notoriously unreliable and are thus not considered further here. Despite this, interview is still the mainstay of personnel selection procedures.

The second method is by 'projective' tests in which the subject is asked to say what he can see in relatively unstructured stimuli (such as 'ink blots'). Such tests are also extremely unreliable and of no relevance to the pilot.

Questionnaires

The third method is by questionnaire, and this method is capable of producing stable and sensible assessments of individuals that relate to their observed behaviour and performance. The basic technique used in the construction of personality tests is known as factor analysis. This procedure initially consists of asking a large population (of thousands) a series of questions about their likes, dislikes, habits, and opinions. When the responses to all of the questions are correlated with one another it is found that patterns emerge in the data. For example, the person who said 'Yes' to 'Do you like driving fast cars?', is also likely to have said 'Yes' to 'Do you like going to noisy parties?' - ie the answers to both of these questions were determined by the same underlying personality characteristic. By following this process of factor analysis, a number of personality dimensions or traits may be identified.

4a.3 Main Dimensions of Personality

At a coarse level, personality may be addressed in terms of just two broad dimensions, which we may term extraversion and anxiety (Figure 4a.1). Extraversion is generally concerned with the traits of impulsivity, boldness,

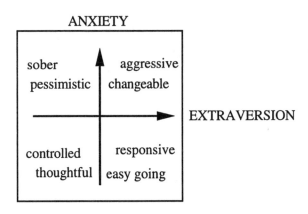

Figure 4a.1. Two-dimensional model of personality

dominance, and sociability. Anxiety is associated with the traits of emotional instability, tension, guilt, and suspiciousness. Some short personality tests resolve personality only into the two broad dimensions, but others are more detailed and may provide scores for a given individual on perhaps 16 traits. The pattern of scores produced by a given individual on all of these traits is sometimes termed the personality profile.

Since extraversion and anxiety are not related to one another (are not correlated), some people will, for example, be anxious and extraverted, and some anxious and introverted. Most people will be around the average in both dimensions (as they are in height, strength, and shoe size), but as the personality departs further from average, so will the characteristics of that personality become more marked. Those who are above the mean, for example, in both anxiety and extraversion will generally be regarded by their friends as aggressive and changeable, whereas stable introverts will be regarded as thoughtful and controlled, anxious introverts as sober and pessimistic, and stable extraverts as responsive and easygoing.

4a.4 Personality-Related Problems of Flying

It may not be surprising that anxious extraverts tend to have more dangerous driving convictions and more flying accidents in which risk-taking plays a part than stable introverts. Anxious introverts, however, may tend to have a rather different sort of accident in which their rigid and sober approach may lead them to perform badly when confronted with an emergency and to mismanage their task when under pressure.

There can be little doubt that risk assessment and risk-taking represent the biggest problem in many accidents, especially those occurring in single pilot operations. Some of these risks may be taken consciously for the purpose of exhilaration (typically, illegal low flying), and some may be taken reluctantly because of commercial pressure. It is also true that pilots may be prepared to take greater risks, and to fly closer to the edges of their ability envelope when they feel that they are being observed - whether this is just by their own passengers or (and more dangerously), if they are demonstrating their aircraft at an airshow.

The average pilot tends to be fairly sanguine (stable and extraverted). Whether this represents the ideal personality type for flying is arguable (and military and civil flying may require different characteristics). Pilots will, however, be able to locate themselves on Figure 4a.1 with some accuracy, and should recognize that if they feel themselves to possess impulsive and sensation-seeking elements in their personality, they should seek to satisfy these personality characteristics away from the flight deck.

The most important effects of personality in civil flying are, however, concerned with determining the ways in which individuals interact with one another on flight decks, and this area of social interaction is now addressed.

4a.5 Social Skills and Interactive Style

There are certain basic ways of behaving that tend to make relationships easy to maintain, whether on or off the flight deck. These behaviours may be divided into verbal and nonverbal behaviours, the latter sometimes being termed 'body language'.

Body Language

The main non-verbal methods of communicating and establishing a relationship are eye contact, facial expression, touch, body orientation and posture, hand and head movements, and physical separation (personal space).

Eye contacts are normally brief. Prolonged staring between those who are not intimate is normally interpreted as threatening and should therefore be avoided. Facial expression is very effective in communicating emotional states – seven of which are sometimes identified; happiness, fear, anger, disgust, contempt, sadness, and interest. Happiness and anger are most readily transmitted in this way, but all facial expressions may be masked, and those of anger, fear, contempt and disgust probably should be on the flight deck. Touch, beyond a handshake, is not a likely form of body language on the flight deck, but posture can be revealing, since we tend to lean towards those with whom we agree and like, but deal with those we dislike by leaning away, avoiding eye contact, and playing with something like a pencil or scratching an ear. The layout of flight decks is fortuitous from this point of view since side by side seating tends to promote co-operation, whereas face to face seating is more confrontational. Leg swinging, finger tapping, and shoulder hunching indicate frustration, disagreement, and tension so should clearly be avoided, for example, by a captain if the first officer is trying to give his analysis of a problem. Personal space invasions are inevitable in the close confines of a flight deck but even here unnecessary physical proximity can cause embarrassment and discomfort and should be avoided. (The classic experiment on personal space invasion was conducted in a urinal. It was found that if a man had somebody standing next to him, the onset of micturation was delayed, and the duration reduced.)

Verbal Behaviour

Verbal behaviours that determine interpersonal relationships include both the manner of speech (pitch, stress, timing, and pauses - paralanguage) and the content. Speed and pitch are important in betraying anxiety, but an important aspect of any conversation is the ease of changeover of speaker. Semantic cues (completion of a meaningful sentence), dropping of voice, eye contact, and gestures are all used to indicate the end of a turn, and the importance of the last two of these is exemplified by the requirement to give a clear verbal cue at the end of your turn in RT conversations. Interruptions, and the ease with which one person gives way to another are very indicative of dominance in the relationship, but as most interruptions break turn-taking rules, they are perceived as rude and domineering, and are thus best avoided. In this context it is interesting to note that Margaret Thatcher, widely regarded as having a dominant style, will continue to speak for up to five seconds when interrupted (ie will speak simultaneously with the interrupter) in TV interviews, whereas most people will give way within half a second in order to avoid simultaneous speech.

Interactive Style

It would be possible to characterize individuals only in terms of their basic personality and the way in which they exhibit social skills, but when individuals are working in a team towards a common goal, it can be helpful to consider the individual's team or interactive style. This is sometimes referred to as leadership style, but on the flight deck what might be termed 'followership style' is just as important in determining the effectiveness of the operation.

A number of ways of characterising interactive style have been suggested, some of which have relevance to pilots. One of these is sometimes termed 'authoritarian'. Such individuals are dogmatic, assertive, and brook no dissent from their subordinates, but, somewhat paradoxically, submit blindly to the authority of those with status and power. While such a 'sergeant major' approach is less common on flight decks than it once was, it is clearly inhibiting and undesirable. The role of dominance in relationships on the flight deck is, however, important, and is returned to below.

The paternalistic style is somewhat different in that the captain will be highly conscious that he is in charge and will wish to see his authority obeyed unquestionningly in flying matters, but will be generous in his praise when the crew carry out his wishes. This style is inevitably frustrating for crew members who will not feel that they are listened to or made effective use of, though the captain will feel that he is trying his best for them. After a well-known near accident, the captain involved described the nature of his relationship with the other crew members as 'relaxed but without undue familiarity', and explained that he had a drill for establishing this relationship. Before the flight he would tell the crew that he was Jesus Christ on the flight deck, but that even Christ needed twelve disciples, and they were his. The captain meant well in saying this, but the effect was not positive.

In the same way that personality may be conceived in a dimensional way, there appear to be two main factors that are important in characterizing interactive style. The first is concern to achieve task goals (goal directed style), and the second is concern to keep team members happy (person directed style). The terms task and emotional leadership are sometimes used with similar meaning.

Figure 4a.2 indicates that we can conceive of the goal and person directed dimensions in the same way as we conceptualized the basic personality dimensions. Again, most of us will be around the middle of this diagram, but those away from the middle will behave in the following ways:

> The P+G- will generally establish friendly relationships on the flight deck, but has too little concern for the task. He will be prepared to leave others to do most of the work, and will resolve conflicts by letting others have their way, even though he may know that this may result in corners being cut and even safety compromised. He is too democratic.

> The P-G+ individual, on the other hand, is so concerned with the efficient conduct of the flight, that he will expect others to behave like automata and will ignore their feelings, thoughts and attitudes. In doing so the P-G+ not only generates an unnecessarily cool atmosphere but actually wastes the expertise of the other crew members, who will not feel inclined to voice their ideas or concerns. He is too autocratic.

> The P-G- individual cares little for either the job or his team members. Such individuals are frequently old pilots who have not gained advance in

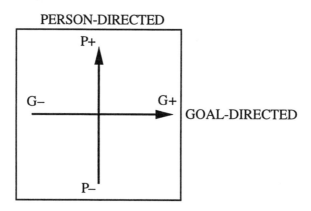

PERSON-DIRECTED

Figure 4a.2. Two-dimensional model of interactive style

their organization and are simply waiting for retirement. This style of interaction generates poorest group performance, with rule bending, failure of group objectives, and poor morale all occurring. These individuals are too laissez-faire.

Ideally, pilots will be P+G+. They will be as concerned for the efficient operation of the flight as they are for the well-being of their colleagues. The P+G+ captain will exercise his powers of command in such a way as to maximize the respect and commitment of other crew members, and will establish an atmosphere that enables all members of the team to feel that their ideas are recognized and considered even if they are not implemented. Some ways of achieving this ideal are discussed below (4a.7).

The style of behaviour that is appropriate may be somewhat context specific, and some of the factors that make for success in leadership may be indefinable. For example, much has been written on why Churchill was a successful wartime, but not peacetime, leader. It may well be that a rather P-G+ style suited war better than peace, not least because in emergency people wish to be given clear orders with confidence. Similarly, on the flight deck, pilots should recognize that a generally democratic approach to problems is desirable as long as time is available, and as long as the democratic approach is directed towards achieving operational goals. When an emergency arises, however, a more autocratic approach may be necessary and unavoidable, but even here it should be ensured that individuals are not too intimidated to speak up if they feel that things are being handled wrongly.

Given that different styles may be appropriate in different contexts, it is generally desirable that a consistency of style is maintained. Those who change from day to day or even minute to minute do not permit their colleagues to develop consistent ways of relating to them, and they will then complain of 'never knowing where you are with him'.

4a.6 Ability, Status, and Role

Ability

The effectiveness of any individual in a team will be determined not only by his interactive style, but also by how competent he is, and how competent other crew members perceive him to be. Perceived competence will interact with his style in determining what other team members generally think of him. For example, the P-G+ individual may be more acceptable to his colleagues if he is extremely competent since they will respect his ability, but if such an autocrat is of low ability he will be regarded with distaste and derision by his team members (whether he is the captain or first officer), and they are likely to dread flying with him. More seriously, the other crew members may knowingly allow such an individual to proceed on an incorrect course of action in the hope that he will wind up in trouble specifically in order to 'bring him down a peg'. The inadvisability of such behaviour on flight decks does not need to be emphasized.

There are more complex effects of perceived ability as well. For example, if a P+G- team member regards one of his colleagues as competent, he is likely to allow that colleague to proceed much further on an inappropriate course of action because he feels that his colleague 'probably knows what he is doing'.

Status

This factor of perceived ability further interacts with status or rank, normally indicated on the flight deck by the amount of gold or silver on the shoulder. It is obvious that a dominant captain will readily question the actions of a junior first officer, but the opposite will not be so. The junior first officer will need to be absolutely certain that the captain really is getting it wrong before he feels compelled to air his anxieties. Even where crew members are of equal status (especially two captains) there may be a good deal of reluctance for one to appear to be questionning the ability of the other, so permitting problems to develop without being questioned or arrested.

Role

A further factor that is plainly instrumental in determining interpersonal behaviour on the flight deck is role. The role of pilots changes to a large degree depending on whether they are the handling or non-handling pilot. It is clear from a number of accidents, however, that it can be very difficult for one pilot to take control away from the other, since doing so may be perceived as a lack of faith in the other's ability.

Three examples serve to illustrate the importance of all of the above factors of style, competence, status, and role and their interaction on flight decks.

a) Two captains were on board a helicopter approaching to land on an oil rig in the sea at night. The handling captain experienced some form of lapse of consciousness just as the aircraft entered the hover, failed to apply power, and the aircraft started to drop towards the sea. The second captain appeared to become alerted to the problem only when the aircraft descended through 100 feet. Even then, however, he did not take control, but instead asked his colleague whether he was all right. Receiving no answer, he again asked him. Only when receiving no reply for the second time did he take control and apply power - too late to stop the aircraft contacting the water but sufficiently early for it safely to fly off. Shortly after leaving the water the pilot who had been forced to take control returned control to the pilot who had been temporarily incapacitated.

b) The twin prop commuter aircraft was commanded by a pilot who was also a senior manager in the airline and known to be somewhat irascible. The first officer was junior in the company and still in his probation period. It was at the end of an already long day, and the captain was plainly annoyed when company operations asked for a further flight, but he reluctantly undertook it. During the approach at the end of this leg, the first officer went through the approach checks but received no response at all from the captain. Rather than question or challenge the captain, the first officer sat tight and let the captain get on with it. The aircraft flew into the ground short of the runway because the first officer did nothing to intervene. It transpired that the captain had failed to respond to the checks not because he was in a bad mood but because he had died during the approach.

c) A twin jet airliner was returning to a major airfield at night. An emergency runway was in use that normally served as a taxiway. A taxiway to the left of this emergency runway was lit in a way that could make it appear like a runway, so the visual scene was rather ambiguous. In the left seat of the aircraft was a first officer who was on his last check ride before becoming a captain. He was pretending to be the captain of the aircraft (but was not the legal commander), and had given away the handling to the occupant of the right seat. This was a training captain, pretending to be the first officer, but who was actually the legal commander of the aircraft examining the actual first officer. The training captain lined up the aircraft with the proper emergency runway and was entirely happy with matters. The first officer, however, was not completely sure that they were lined up with the proper runway for a number of reasons, and felt compelled to say something. Not wishing to appear as though he was unclear about what was going on, however, he phrased his question in a very non-committal way, and said to the commander 'You are going for the emergency runway aren't you?'. The captain, also wishing not to betray any incompetence replied 'Yes, of course I am', but actually interpreted the first officer's remark to mean that he was going for the wrong runway. While the first officer had his head down, the captain altered his approach, landed on the taxiway, and was fortunate to stop before colliding with a taxying aircraft.

4a.7 Group Decision Making

One reason for having more than one person on a flight deck is to produce better quality decisions and better solutions to problems than would be produced by just one. It is generally true that the decision made by a group will be better in quality than the average decisions made by the members of the group, but perhaps slightly depressing that group problem solving will rarely improve upon the problem solving ability of the ablest member of the group. From this point of view, therefore, the function of having more than one person in a crew is to improve the chances of having an able person there, rather than to have an interaction between crew members that produces better decisions than any would produce individually.

Recognizing and accepting the correct solutions and decisions within the team is a process affected by a number of factors. These include the ideas of conformity, compliance, status, risky shift, and group duration.

Conformity

The classic experiment that demonstrates the power of conformity is one in which a subject is asked to judge the length of some lines drawn on a piece of paper and to compare them with the length of a standard reference line. The correct answer is reasonably obvious, but the subject is placed in a group of stooges who give their answers before him. Initially, all the stooges give the right answer, but as the experiment progresses the stooges give the same, but obviously wrong, answer. When it comes for the subject to give his answer, he is likely to give the same answer as the stooges even though he really believes it to be wrong. An interesting aspect of this experiment is that the effect is almost maximized when the size of the group holding the opposing opinion is just four.

Compliance

Compliance is the term used to describe the individual's likelihood of complying with a request. If a large and unreasonable request is made, there is a greater likelihood of it being complied with if it has been preceded either by an even more outrageous request that has been denied, or if it has been preceded by a smaller, more reasonable request that has been accepted. In a field experiment, for example, it was found that a householder would be more likely to agree to having a large road safety poster in his front garden if he had either previously refused to have an even larger sign, or had accepted having a smaller sign.

Status

The role of status in group decision making is also of clear importance. The following (trivial?) problem was given to USAF bomber crews: 'A man bought a horse for $60 and sold it for $70. He then bought it back for $80 and sold it again for $90. How much money did he make in the horse business?' The correct answer to the problem was computed by 30 per cent of the high status pilots, by 50 per cent of the medium status navigators, and by 30 per cent of the low status gunners. However, over 90 per cent of pilots who got the right answer persuaded their group of the answer, but only 80 per cent of the navs and 60 per cent of the gunners managed to do so.

Risky Shift

If a group is asked to contemplate a problem such as whether an individual should give up a secure but modestly paid job for a less secure but better paid job, they will usually come to a decision that is more risky than the average made by individual group members. In fact, this is an example of a more general phenomenon that might better be termed polarization. If the individual members of a group already tend to an attitude or view, then the group will tend to that view even more strongly. The problem this creates on flight decks is that many pilots like to be thought of as fairly bold individuals, and combining a set of such individuals into a crew can make for an unduly bold outcome.

The first and third of these factors (and some of the points made in 4a.6) are exemplified in the following, slightly paraphrased, extract from a confidential incident report made to the UK CHIRP system (cf ASRS in USA). A three man crew in a modern aircraft had just made two missed approaches but had seen the runway lights on each occasion.

> As we entered the hold to contemplate the situation, another aircraft landed. A suggestion was made that we knock 50 feet off the decision height, but as this was not legal, it was ruled out. We then managed to delude ourselves that flying level at the DH of 220 feet was a legal and reasonable way to achieve our landing... The P2 flew a manual   approach to 220 feet and levelled off. Within seconds the captain   shouted 'I have control' and the aircraft continued to fly more or less level for several seconds. From my position I studied the information with a growing feeling of unease. The glideslope was fully fly down and had been for some time, we had no way of knowing how far we were beyond the threshold. The brightness of the runway lights convinced me that the captain could see and knew what he was doing. I called out our radio height at the same time as the equally concerned P2 called the speed and rate of descent. The next few seconds saw the cockpit filled with height, speed and rate of descent. We touched heavily... followed by a 'max performance' stop. We cleared the runway well down. The pregnant silence which followed served to reinforce the feeling that we'd been party to an act of supreme folly and bravado and were lucky to escape with a few grey hairs and severely battered personal pride.

Group Duration

A final factor that can affect the way in which a group makes decisions is the length of time that the group has existed. In military flying, the flight deck team may fly together regularly (the 'constituted' crew) and come to know one another's habits, but in civil flying such an approach would be impossible and the crew may well be strangers at the beginning of a flight. This reinforces the need for standardized procedures since they serve to enable each of these strangers to know what the other members of the crew will do. Although it may be thought preferable to maintain a group whose members know one another, there are large disadvantages to this approach. If the crew comprises members who are incompatible with one another, this 'bad' crew will stay together for a long time and relationships may deteriorate to a dangerous extent before some crew shuffling is undertaken. It is also possible that the crew which operates effectively comes to rely more on interpersonal knowledge than on adherence to standard procedures, with potentially serious consequences.

4a.8 Improving Group Decision Making

In one study of group decision making, two groups were given the same problem to solve, but the members of one group were given the following guidelines:

a) Avoid arguing for your own individual judgements. Approach the task on the basis of logic.

b) Avoid changing your mind only in order to reach agreement and avoid conflict. Support only solutions with which you are able to agree at least partially.

c) Avoid conflict-reducing techniques such as the majority vote, averaging, or trading, in reaching your decison.

d) View differences of opinion as helpful rather than a hindrance in decision making.

The group that received the guidelines produced the better performance. This strongly implies that individuals may be trained in the group behaviour that helps in decision making. Aviation has traditionally been very strong in training individual skills and rule based behaviours (2b.6 & 2b.7) but has not generally provided pilots with practice at solving possibly ill defined problems on a group basis. It is this deficit which Line-Oriented Flying Training (LOFT), Flight Deck Management (FDM), and Cockpit Resource Management (CRM) training seek to redress.

Such training often includes video taping realistic simulation exercises. This is aimed to provide pilots with 'behavioural feedback' so that they can observe themselves, and possibly realise that the way that they come across to others is rather different from the way they intend. Thus their self-image is made more consistent with the image others have of them.

In such exercises, pilots also may be required to 'role play' or act in a certain way. The object of this is to require perhaps the timid first officer to be more assertive in a benign environment both so he will realize that he is capable of behaving in that way, and so that he will appreciate that most captains will respond favourably to being provided with a clear statement of his ideas. It may also be useful to require the authoritarian captain to demonstrate to himself that asking other crew members for their advice and ideas will not be seen by them as a sign of his weakness, but that doing so serves to consolidate the crew and lead to better decision-making.

Both of the above may be regarded as techniques for training social skills, but there are also guidelines that can be provided in leadership and followership. The most important principles for the leader to follow in order to come to a good group decision and maintain the morale of the team are:

a) Avoid giving any indication of his own opinion or ideas at the outset. If any member of the team has another idea, he will then be reluctant to air it since it will appear to be contradicting the captain.

b) Specifically and overtly solicit the ideas and opinions of crew members and, especially, encourage them to express doubts and objections to a particular course of action. Always ensure that the potential problems of a course of action are fully aired and not ignored.

c) When the leader has made a decision, he should explain to the other crew members his reasons for arriving at that decision. Failure to do so will make the crew members feel that their own ideas have either not been considered or even heard.

There are many other guidelines that may be given that are as appropriate for leader or follower. The following are worth bearing in mind:

Don't delay airing your uncertainties or anxieties because you are worried about looking foolish or weak. Other crew members may well be feeling the same as you and will welcome some candour.

When your opinions or ideas are sought, give your point of view fully and clearly, without worrying about whether you are saying what the other person wants to hear - but don't do so in an emotionally loaded or unnecssarily dominant way (eg 'Any fool can see that...').

Don't become 'ego involved' with your own point of view and simply try to get your own way; deal in evidence and not prejudice. If a group decision has been made, accept it unless you feel that it contains some hazard not appreciated by the other group members.

Don't let others progress down wrong paths of action and get themselves into trouble just to make yourself look clever.

Don't compete, don't get angry, don't shout, and don't sulk on the flight deck. Don't let your own bad mood show. Do what you can to maintain a pleasant working atmosphere even if you don't much like the other crew members.

4a.9 Interacting with ATC, Cabin Crew, and Passengers

People are very good at identifying themselves with a group, but the pilot is a member of more than one group. All of the foregoing has considered him as a member of the flight deck team, and it is clearly with this group that he has most identity of purpose and interests in common, but it should briefly be recognized that there are other groups of which he is also a member. The first is the total crew of the aircraft.

It is slightly unfortunate that in identifying with a group, (the 'in' group), people also identify those who are not members of that group (the 'out' group). Under some circumstances the cabin crew will clearly regard the pilots as an out group and thus emphasize their negative characteristics (lazy, overpaid, prima donnas), and the pilots will do the same for the cabin crew. If faced with an aircraft full of young, drunken holidaymakers, however, the cabin and technical crew rapidly become one 'in' group with the passengers as the 'out' group.

Under other circumstances (such as a delay caused by an ATC strike), then passengers and crew become one group. None of this is likely to cause difficulty except under special circumstances. For example, if a member of one group has such a poor opinion or lack of contact with another, he may fail to make use of the members of that other group when he should. The technical crew are not able to see what is going on in the cabin, but may be so unused to the cabin crew giving them information about events there (perhaps smoke or strange noises) that they fail to use them when they could. Conversely, the cabin crew may know so little about what happens on the flight deck that they do not realize how much help they can be.

When things do not go well, the pilot should endeavour to bear in mind that ATC, cabin crew, maintainers, and passengers all have an identity of purpose, and that little is to be gained by treating any of these groups as though they are the enemy.

4b DESIGN OF FLIGHT DECKS, DOCUMENTATION, AND PROCEDURES

Introduction

The guiding principle of flight deck design is that it should be accomplished in a way that fits the job to the man rather than the man to the job. Unfortunately, since humans are flexible in a variety of ways, there has always been a temptation to permit poor, but possibly financially economical, design on the flight deck to enter service in the expectation that the pilot will be able to cope. He may well be able to do so, but with a possible cost in terms of increased likelihood of error, discomfort, and fatigue.

Although few pilots are likely to be able to influence or change the design of their flight deck, some knowledge of design principles may allow the pilot to carry out his duties in the greatest safety and comfort.

4b.1 Workspace and Comfort

Anthropometry

It is obvious that the most important factor that influences the overall size of aircraft and their flight decks is the size of people. People, however, vary greatly in size from one another (the average Central African Pygmy is 1.37m tall, whereas there is a Sudanese tribe whose members average 1.83m - over one third taller), and the aircraft flight deck is required to cope with all - or at least most - of this size variation.

The study of human measurement (anthropometry) has resulted in a great deal of data being available to designers on various aspects of the sizes of different populations. These data are sometimes divided into *static* measures such as joint to joint (eg ankle to knee) or *contour* (surface measures, eg from finger tip to elbow) and *dynamic* measures such as reach and limb clearance envelopes. Given these data, the designer must decide how great a proportion of the population he wishes to accommodate on his flight deck. A common decision is that all those between the 5th and 95th percentiles in a defined population must be catered for. For example, the 5th percentile of height for British males is 1.625m (ie 5% of British males are shorter than this) and the 95th, 1.855m. The comparable figures for women, however, are 1.505m and 1.710m; this simple example serves to emphasize the importance of considering the appropriate population before making design decisions.

Eye Datum

A basic feature of the design of a flight deck is that the pilot should be able to view all important displays and be able to maintain an adequate view of the outside world without the necessity for more than minimal head movement. Thus the cockpit space must be designed around a defined position of the pilot's eye (sometimes termed the eye datum, reference eye point, or design eye position). This position is often made apparent to the pilot by the provision of two small balls on the central windscreen pillar which appear aligned only when the pilot's eye is at the designed point.

External view is of particular importance in the definition of the design eye position. The pilot must, without strain, be able to look down over the top of the instrument panel and see sufficient of the ground ahead to enable him to land the aircraft. One standard requires the design eye position to enable the pilot to be able to see a length of approach lights that would be covered in three seconds during an approach.

Once the design eye position has been set, and the anthropometric range of pilots decided upon, the size of the cockpit workspace and the degree of adjustment of seat, rudder pedals (and possibly other controls) will follow. This process may, however, still be constrained by other factors: for example, the cockpit is inevitably placed at an aerodynamically important location that may limit its width, and the size of cockpit windows may be limited by cost and their weakening effect on fuselage structure.

It is important for the pilot, however, that not only are all of his controls within reach, but that he is able to sit comfortably for considerable periods of time. For this reason, a great deal of effort has gone into the design of pilots' seats, but these still require the pilot to adjust them properly. Many studies have shown the high prevalence of lower back pain and discomfort in the population in general and in pilots in particular. The short cylinders of bone (the vertebrae) of which the spine is composed are separated by shock absorbing discs. The function of the seat's lumbar support is to push the lower spine into a shape that causes the compression loads to which the discs are subjected to be distributed evenly across each one. If the lumbar support is removed, allowing the spine to bow outwards, the asymmetric load on each disc causes it to bulge. This bulge in turn causes the muscles and ligaments in the area to be put into tension to relieve the load on the disc, resulting in backache. For these reasons it is important that the pilot's seat should be in good condition, and that the pilot should take care both to adjust it properly and to sit in it with the pelvis pushed back to enable the lumbar support to do its work.

4b.2 Displays

Perhaps the most important requirement in display and control design is that of standardization. This enables the pilot who is accustomed to one particular organization and manner of display to be able to transfer his experience to other individual aircraft and other aircraft types without confusion or 'negative transfer'. Total standardization is not possible for a number of reasons, and might not enable new developments in display technology to be taken up; for example, the best way of presenting altitude information on a conventional round gauge may not be the best way of doing it on a cathode ray tube (CRT) in a 'glass' cockpit.

Flight Displays

Most aircraft with conventional instruments have used the 'T' type of arrangement of the basic flying instruments illustrated in Figure 4b.1. This co-locates the most important flight displays around the highest priority instrument – the artificial horizon or attitude director. To some extent, the spatial relationship of airspeed, attitude, and altitude has been maintained even

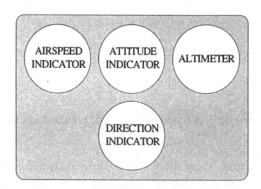

Figure 4b.1. The standard 'T'

in glass cockpits, but with revised scale design. In such aircraft, the attitude display appears traditional with a vertical moving strip or tape display of airspeed to its left and a similar scale displaying altitude to its right – see Figure 4b.2. Although it has already been emphasized that standardization of display organization is highly desirable, there are arguments for presenting speed information on such tape displays with the large values at either the top or bottom of the display. Generally, large values are expected at the top of a display, but if this practice is adopted on the vertical scale of airspeed, it means that, when the aircraft is climbing at a fixed power setting, the speed and altitude scales will be moving in opposite directions (as the aircraft climbs and slows), possibly producing a perception of roll in the pilot. It also means that as the aircraft slows, the tape rises, a practice that may go counter to population expectations. Although these arguments may seem contrived, they have resulted in two different types of speed scales that move in opposite senses to one another in modern aircraft; one has large values at the top of the tape, and the other is vice versa. Such a state of affairs is plainly undesirable, and has occurred because there is presently no requirement for objective evaluation of the human engineering acceptability of instrumentation, and no mechanism for standardization.

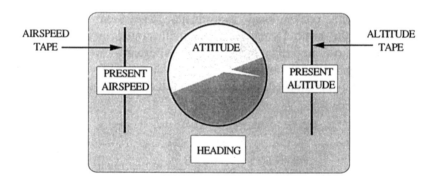

Figure 4b.2. A typical 'glass cockpit' primary flight display

In the above example, there may seem little difference between presenting altitude information on a dial type instrument or on a moving scale, but there are strengths and weaknesses in various display designs. A classic example of this concerns the three-pointer altimeter (Figure 4b.3). Such instruments are commonplace, yet straightforward laboratory evaluation shows that a three-pointer altimeter takes about 3.9s to read compared with 1.3s for a digital instrument, and up about one fifth of readings can be in error compared with none for the digital instrument. Multi-pointer instruments have been known to be difficult to read for many years, but the analogue nature of single-pointer displays can sometimes be an advantage over digital presentation. For example, although digital altimeters are probably advantageous in some respects since the pilot conceives and deals with his altitude or flight level in terms of a numerical value, analogue displays of radar altitude may be preferable for the helicopter pilot who can make a rapid glance inside his cockpit and obtain a good idea of his height above a surface from the position of a single pointer on a scale.

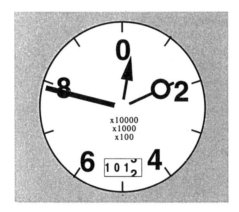

Figure 4b.3. The three-pointer altimeter

This assumes, of course, that the scale is stationary with a moving pointer or marker. Moving scales with out-of-sight end points compel the value of the scale to be read and do not provide analogue information. A method of displaying altitude (and other) information that is now available is digitized voice. Although such a display would be obviously inappropriate during normal flight, a voice presentation of radar altitude during the final stages of an approach may be of great benefit to the pilot, preventing the necessity to look away from the runway. An alternative way of achieving the same goal is by the use of a head-up display (HUD). By projecting collimated (focused at infinity) information onto an angled glass through which the pilot observes the outside world, the pilot is enabled to acquire flight information without having to divert his gaze. Such displays are now entering civil use, and undoubtedly confer benefits during poor weather approaches. It is important to bear in mind, however, that although HUDs do not require the pilot to make large changes in the direction of his gaze or in the focus of his eyes, they do necessitate an attentional shift from outside world to symbology in order for the flight data to be perceived.

Plainly, the type of display used should be determined by the use to which the pilot is to put the information, and the situation in which it is to be used. The designer should thus have addressed a number of basic questions. Will the pilot require purely quantitative data from the display? - if so, a digital presentation should probably be used. If, however, qualitative and rate information are more important, then some form of analogue display is probably preferable. If reference to the absolute end point of the scale may be important, then moving tape displays should be avoided. Lastly, if the display is to be used only in order to enable the pilot to know when some parameter (eg engine vibration) enters an undesirable region, then a simple warning (see below) is preferable to either analogue or digital displays.

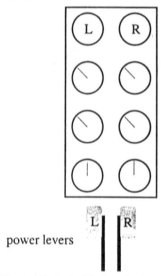

power levers

Figure 4b.4. An ideal engine instrument layout

The organization of displays on the flight deck has already been discussed briefly with regard to the layout of the basic flying instruments, but the layout of engine instruments is of equal importance. Figure 4b.4 shows a conventional layout of such instruments arranged so that the columns contain instruments associated with a particular engine, and the rows contain instruments of one type. Such an arrangement maximizes the likelihood that disparate readings on, say, the N1 gauges will be noticed, and also maximizes the probability that the pilot will associate the left engine instruments with the left engine and not, for example, shut down the wrong engine. The situation is less straightforward, however, if secondary engine information (such as oil temperatures and pressures) must be displayed. Figure 4b.5 illustrates two different solutions to this problem. It is arguably likely that arrangement A makes disparate readings on the secondary instruments hard to identify, but has the strength of keeping the instruments for each engine clumped together and thus reduces the likelihood of the wrong engine being shut down. The opposite could be argued for arrangement B. Both of these arrangements, as well as some others, exist in current aircraft, though, doubtless, careful evaluation would reveal an optimal solution.

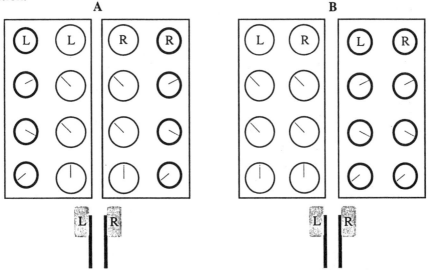

Figure 4b.5. Alternative layout of primary and secondary engine instruments

Flight decks should be illuminated by a combination of even floodlighting and internal display illumination. This should enable the displays to be read easily, permit reference to cockpit documentation, and allow controls to be seen and identified. Adjustment should be available to allow for variations in both personal taste and ambient light levels. Harsh shadows, glare, and reflections should be avoided.

4b.3 Controls

Whereas displays are devices for enabling information to pass from the aircraft to the pilot, controls enable information transfer from pilot to aircraft. There are certain basic considerations, mostly obvious but frequently ignored, that govern the way in which controls should be designed and arranged.

a) Frequency of use - Controls should be located such that they are within the reach envelopes of all designed users of the aircraft. Furthermore, controls that are used frequently or for protracted periods should be located so that they do not require an awkward or fatiguing posture in the pilot.

b) Sequence of use – Controls that are frequently used in a given order should be laid out so that the sequence of use is represented in the layout of the controls. This is important not only because it makes their use convenient, but also because the layout itself then acts as a prompt for the pilot.

c) Importance of control – Important controls such as the throttles and flap should be located in easily reached and unobstructed locations.

d) Simultaneous use – Certain controls (eg throttle and trim) may, ideally, be operated simultaneously. They should therefore be located to enable simultaneous operation. This consideration has led, in some military aircraft to the HOTAS (hands on throttle and stick) concept by which all secondary controls (eg flap, trim, transmit) are located on the two primary controls.

e) Visual/tactile dissimilarity – Many aircraft contain rows of switches that control different functions. Although it may be convenient and aesthetically pleasing for such switches to be identical to one another, making them dissimilar reduces the likelihood of inadvertent operation.

f) Symbolism in control design – Many controls have been designed to contain some reference to their function. Thus undercarriage levers are often provided with a handle that mimics a wheel, and flap levers may resemble a cross section of flap.

g) Control/display compatibility – Controls should be located such that they maintain some sort of spatial logic with the display that they are associated with. Thus the columns of engine instruments in Figure 4b.4 should be aligned with their relevant throttles and start levers.

h) Control loading – The force required to operate a control should not be within that which can be exerted by the target population of pilots, but should be harmonized with the forces required by other related controls. For example, a control column will be difficult to use if it requires large forces to control roll but demands only light forces to control pitch.

i) Prevention of inadvertent operation – Controls should be designed to minimize the likelihood of inadvertent operation. Where this could be hazardous, the control should be fitted with a guard.

j) Standardization – Last, but most important, controls should be standardized in their location and sense of use from one aircraft to another, and between different aircraft types.

In practice, not all of the above requirements can be met, but some forms of shortcoming should not be required as tolerable. The following CHIRP reports illustrate some of these problems.

> For the third time, I was caught by variation of switch position on our F27s. During the after start checks, the F/O put the water/methanol switches on instead of the pitot heaters. On some aircraft these switch positions are exchanged. As full power was achieved, I was surprised to hear water–methanol flow cutting in (my own taxy checks having failed to spot the ergonomically induced error)... I know that others have made this error several times, though not usually reaching the take-off stage.

> On intermediate approach, while moving his hand from the VHF frequency selector switch on the central pedestal to the heading select knob on the glareshield, the captain's right knuckle contacted the go-around button on the left thrust lever - with the expected result.

> ...while I set the QFE and promptly commenced the descent to 2 000ft QFE... At 2 000ft my co-pilot said 'You've gone below 2 000ft'. I replied that I had not, but then saw that my altimeter was set on 1030mb and not the correct QFE of 1020mb. Consider the attached diagram (Figure 4b.6), drawn actual size. The altimeters are viewed from a distance of some 50cm, while the instrument panel is acknowledged to suffer from shake. The individual helicopters are fitted with altimeters of types A and C, or B and C. As the pilots fly from either seat, a pilot may find himself using an instrument of any type. It seems that most of my colleagues have difficulty in seeing and setting the correct pressures.

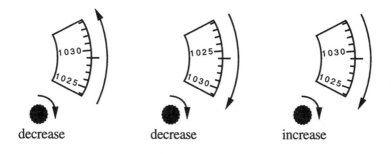

Figure 4b.6. Altimeter control–display relationships

Doom; back to work, first trip, a recency check. Shortly after take-off, captain requests engine antice 'off'. Despite nearly seven years experience on the aircraft, I put the adjacent hydraulic pumps off, causing amber warning lights and muttered curses from the captain. Although my company has tried to differentiate between the four similar switches by removing the white rubber covers from both the antice switches, this problem still occurs. We have been asking for guarded toggle switches for the hydraulic pumps for years and still no luck.

4b.4 Warnings

It is essential that any warning should be 'attention–getting', but that it should perform this alerting function without being startling. A principal problem for the designer is deciding what is an important enough problem to warrant attracting the pilot's attention to it. Several accidents illustrate that this decision is not always made correctly. For example, the Trident did not warn the pilot of early 'droop' (slat) retraction before the Staines (1972) accident, and the 737 contains no warning of excessive engine vibration even though this was a central feature of the Kegworth accident in 1989. Both these aircraft, however, provided good displays of the state of these variables.

A second function of a warning is to inform or report (since there is little point in alerting the pilot without telling him what is wrong), and it is sometimes suggested that it is also important for warnings to guide the pilot to the required actions.

The alerting function for all important failures should be fulfilled by a audio warning for the obvious reason that even the most conspicuous visual warnings rely on head and gaze orientation. Accidents have occurred to aircraft such as the Jaguar, which has no audio warning for undercarriage, because the pilot has landed, gear up, with a large flashing light in the gear lever that has gone unnoticed.

Perhaps an ideal organization for an aircraft warning system is for a single audio warning to alert the pilot to any failure and to direct his attention to a single central warning panel that announces the nature of the problem with a suitable illuminated caption. Such a sytem uses audition and vision appropriately.

Unfortunately, a common problem in warning systems is that designers have used a number of abstract audio warnings to mean different things – thus a bell may indicate fire, a horn for gear, and so on. Some aircraft even use the same sound to mean different things depending on whether the aircraft is on the ground or in flight. The problems with such an approach are that the pilot may well not recognize the abstract audio and be compelled to interrogate the annunciator panel anyway, or, more seriously, that he may wrongly identify the audio and thus deal with the wrong problem. The provision of voice warnings may not cure this problem since if the voice message is made only once, it is easy for the pilot (possibly already attending to a visual problem) to miss it, and repeated voice warnings can be irritating and disruptive of communications.

There are a number of recorded instances of pilots misidentifying abstract or coded audios, but perhaps the most spectacular example of misinterpretation of an abstract warning occurred in a DC10 (1979). This aircraft is fitted with a stick shaker to warn of the impending stall. The crew in question, however, believed that they had an engine problem so that when the aircraft approached the stall and the stick shaker actuated this was interpreted as severe engine vibration and an engine was shut down. Naturally, the aircraft then stalled, lost a great deal of height, and the crew was fortunate to recover the situation before hitting the ground.

Lastly, it is important for warning systems to be reliable in the sense that they respond to all genuine problems, but do not generate false alarms. Ground proximity warning systems (GPWSs) were well known for generating spurious warnings when first introduced, and it has been suggested that accidents have been caused by pilots (used to hearing spurious warnings) ignoring genuine instances.

The only general advice that may be given to pilots with regard to warnings is that all must be taken seriously, but none acted upon until the pilot is certain of the nature of the problem.

4b.5 Checklists and Manuals

Pilots are more likely to be the victims, rather than the perpetrators, of poorly designed manuals and checklists. An often cited example of the importance of good design in this area concerns the Tri-Star accident at Riyadh (1980) in which over 300 people died. A cargo fire had compelled a forced landing, which was completed safely. After the aircraft came to a stop, however, the crew spent three minutes looking through their documentation for the procedure to deal with an aft cargo smoke warning, and it was during this time that passengers and crew died from toxic fumes and fire.

This incident highlights the requirement for good accessibility of information in manuals and checklists. Although there is plainly a requirement for crews to be sufficiently familiar with their documentation that they will know where to find relevant information, it is also essential that the fullest use is made of good, cross-referenced indexing, colour coding of pages by topic, and dividing pages with protruding thumb-locators.

Manuals and checklists should be kept to a minimum in size in order to make them easy to use on the flight deck, but, at the same time, text size should be kept well above the minimum required for bare legibility since they may well have to be read under poor lighting conditions by a presbyopic (see 1b.2) crew that already has a high workload.

It is also important that the amount of information included in documentation is relevant to the needs of the pilot, that it is presented in easily understood language, and in a type face that maximizes legibility. Upper case text and italics may be useful in conveying emphasis, but it is interesting to note that neither of these are as fast to read as normal text. THUS LONG MESSAGES, IN UPPER CASE, SUCH AS THIS, SHOULD BE AVOIDED SINCE WORD SHAPE - WHICH ACTS AS A CUE IN READING - IS ESSENTIALLY LOST. Colour may be a preferable way of categorizing information and giving importance to different sections of text, but the legibility of different text/background combinations varies widely, and, for example, red text on a white background may become effectively invisible under red light.

In using checklists it is important for the pilot to discipline himself to adhere to the designed procedure. If the checklist calls for a challenge and response, then this is the way in which it should be used. Two common problems arise in the use of checklists. The first is that items may be omitted, often because the progress of the checklist is interrupted by an external event (eg copying a clearance), or simply because a pilot, using his thumb as a marker, adjusts his grip on the checklist and misses an item. Some aircraft have been fitted with mechanical checklists for important items which enable sliding windows to cover items that have been actioned, and checklists have also been produced in plastic that enable completed items to be marked with erasable pen. In the absence of such aids, the pilot should be aware of the ease with which checklist items can be missed, and take special care when resuming an interrupted check.

The second major source of error in using routine checklists is that they may be responded to automatically rather than diligently. It is tempting for the pilot to regard a rapid dismissal of checklist items as indicative of his skill and familiarity with the aircraft, but, if checklists are dealt with in this automatic way, it is very easy for the pilot to see what he expects to see rather than what is there (see 2b.1). Although it may be difficult to devote care to a procedure that has become routine, this is exactly what the pilot is required to do.

4b.6 The Glass Cockpit

It is generally true to say that before the introduction of computers into commercial aircraft, cockpits were inevitably complex because every sensor in the aircraft (whether of airspeed, oil quantity, or cabin altitude) was connected to its own display on the flight deck, and the value of that parameter was thus displayed constantly. The two major changes that the automation has enabled are that the computer is able to receive information from many data sources and integrate it into a single comprehensive display, and that the computer can also be selective with regard to the amount and type of information that is displayed at any given time.

The best example of the integration of information is in the navigation or horizontal situation displays found in advanced flight decks. These take information from many data sources such as ground radio aids, the aircraft's inertial platform, and weather and ground mapping radar, to present a single integrated picture to the pilot of all available information relating to the two horizontal dimensions. The pilot is thus freed not only from the requirement to integrate the information from these data sources for himself, but is also freed of many inferential tasks (such as calculating wind strength and direction) that are also carried out by the computer.

Ideally, being relieved of tasks that can be automated should leave the pilot better able to make the higher level decisions that only he can make, and give him much improved 'situational awareness'. This is certainly so, but it is also possible that by providing the pilot with such an attractive and compelling display, he may have been distanced from the real world and tend to believe exactly what the display tells him rather than regarding it as information from which to build an internal model (see 2b.1) of the real world.

The selective nature of displayed information in glass cockpits is exemplified by current aircraft status displays which provide information of, eg, control surfaces, wheel temperatures and pressures, or cabin temperature and pressure, automatically tailored to the pilot's activity and the phase of flight. Although this approach undoubtedly acts further to reduce pilot workload, there is the possibility of the pilot remaining unaware of important information when solving an unusual and unexpected problem.

CRT displays also enable the wide-ranging and flexible of colour, but this facility should normally be employed with restraint. Colour is useful in all types of display for categorizing information and some standards and conventions already exist in this regard. Flashing red should be used only for information requiring immediate attention, red and yellow or amber for less immediate problems, and white and green for satisfactory or non-critical information. In the glass cockpit, it is possible for the colour of symbology to change to indicate a change of state, for example, from 'ALT (altitude) capture' (blue) to 'ALT hold' (green). Such colour changes are normally useful, but flight decks should not be designed so that the state of a variable is indicated only by colour with no associated change of caption text or location.

Although the term 'glass cockpit' derives from the use of cathode ray tube (CRT) displays, such aircraft also tend to contain automation of controls. Since such flying controls may be limited in their authority by the aircraft's computers to a safe flying envelope (eg making the aircraft impossible to stall), there is a danger that the pilot may come to regard the aircraft as infallible, and able to cope with impossible situations. It is no use, for example, to be flying an unstallable aircraft if the pilot has placed it at low level and low airspeed, with no energy available to fly away from the problem. Automated aircraft do not, therefore, absolve the pilot from operating in a way that complies with the basic requirements of safe flight.

Other problems perceived by the pilots of 'glass cockpit' aircraft are that the displays are so easy to use that they may make it difficult when they fail for the pilot to use his traditional skills at basic instrument flight, and that this might be especially true for young pilots who do not have any depth of experience on more basic aircraft. Another concern is that the complex systems which drive the modern pilot/equipment interface cannot be understood by pilots to the same extent that more basic systems could, partly as a direct result of the complexity, and partly as a result of 'need to know' mechanized teaching methods. It is for these reasons that the pilots of such aircraft are supposed always to be saying either 'What's it doing now?' or 'I've never seen it do that before'.

Lastly, but perhaps most importantly, glass cockpits do appear to produce the problem of what might be termed 'mode awareness'. Since the automatic flight and engine management sytems can be set up in so many 'modes', it is possible for the pilot to believe that the aircraft is programmed to carry out one function when it is, in fact, performing another.

It is plainly important for the pilots of such aircraft to maintain an accurate knowledge of the aircraft's status by including the mode representation as a central part of their scan, and it is important for designers to ensure that mode information is always centrally and clearly displayed.

4b.7 Intelligent Flight Decks

There is no precise line that divides the 'automated' from the 'intelligent', but the sophistication of the data evaluation and problem solving of which modern aircraft are capable would appear to merit the use of terms such as 'pilot's associate' and 'electronic crew member'. There are probably three main human factors issues, presently unresolved, that may be identified.

The first concerns the level of autonomy given to the machine. Should the computer be permitted to evaluate data, make decisions, and execute them without reference to the pilot, or should it take a more advisory role, presenting suggestions to the pilot to help him to make the decisions? For example, should an aircraft fitted with a traffic collision avoidance system be permitted to alter the aircraft's flight path if it detects a potential conflict? Many such problems will be solved individually, but it is widely felt that a philosophy should be developed of how the pilot and machine should interact, and to decide whether future aircraft should be 'human centred' or 'automation centred'.

The second concerns the representation of uncertainty. As machines become more sophisticated and as they evaluate greater quantities of possibly 'noisy' data, so do the solutions they arrive at become more 'probabilistic' and less 'deterministic'. Presently, aircraft displays do not tend to give the pilot any estimate of the reliability of the data displayed, they simply display the machine's best guess. An example of this is the navigation display in glass cockpits. The aircraft's position is computed on the basis of inertial navigation and reference to ground based aids, but more or less the same display is given to the pilot regardless of whether the aircraft 'knows' that good data is being received from all sources or whether it 'knows' that it is receiving information from one, poor quality, distant DME and an inertial system that may have been drifting for four hours.

The third issue stems from the two above and concerns 'trust'. The pilot obviously needs to have an appropriate level of trust in his equipment since overtrust has obvious dangers and undertrust can lead to unnecessary workload and operational difficulties. The operational reliability of systems is plainly an important determinant of the level of trust that pilots have in their equipment, but it is also possible that modern displays may be so compelling that they engender more trust in them than they actually merit.

4b.8 Boredom, Situational Awareness, and Maintaining Skills

Highly automated flight decks and extended range operations have developed more or less concurrently. This means that the cruise phase of flight leaves the pilot with little to do, but may continue for as much as 14 hours. The problems of both boredom and loss of handling skills are apparent. In such long haul operations, the problems may be exacerbated by the requirement to carry two crews in order to comply with flight duty time regulations.

It is plainly unsatisfactory for crews to become bored since this may lead not only to reduced monitoring of their environment and consequently reduced situational awareness, but may even lead to their falling asleep. Even more serious is the possibility that a bored crew may be tempted to experiment with systems on the flight deck. This has already led to at least one serious incident in which the crew, attempting to discover how certain aspects of the aircraft's autothrottle operated, disabled the engine management system to the extent that fan blades were shed by an engine, penetrated the fuselage, and a passenger was lost through the hole.

Although many suggestions have been made for ameliorating these problems (and there is no doubt that pilots must, presently, do their best to manage the flight in a way that enables them to stay as alert as possible, and companies must organize their training so that basic handling skills may be maintained), the aircraft of the future will demand a radical reassessment of the pilot's task. It must be hoped that pilots, designers, and human factors specialists will produce a flight deck that is not only safe, comfortable, and efficient, but one that provides its occupants with the stimulation and interest that has always made flying so worthwhile, and pilots so committed.

FURTHER READING

Borbely, A., (1986), *Secrets of Sleep*, Basic Books Inc., New York, USA.

British Airline Pilots' Association Medical Study Group, (1988), *Fit to Fly: A Medical Handbook for Pilots*, 2nd Edition, BSP Professional Books, Oxford, UK.

Civil Aviation Authority, (1990), *CAP 567 - Aviation Medicine Manual*, Civil Aviation Authority Publications, Cheltenham, UK.

Civil Aviation Authority, Aeronautical Information Circulars:-

AIC 95/1991 Effect of flickering light on helicopter passengers and crew.

AIC 105/1992 Pilots and spectacles.

AIC 115/1992 Skin contact with aviation fuels.

AIC 127 1992 Blood, plasma and bone marrow donation.

AIC 132/1992 Hypoxia in flight and its prevention.

AIC 16/1993 Medication, alcohol and flying.

AIC 2/1995 Malaria.

Ernsting, J. & King, P., (1988), *Aviation Medicine,* 2nd Edition, Butterworths, London, UK.

Funk, C.S., (1995), *Human Factors in Flight: Instructors Guide,* Avebury Aviation, Aldershot, UK / Brookfield, USA.

Funk, C.S., (1995), *Human Factors in Flight: Student Workbook,* Avebury Aviation, Aldershot, UK / Brookfield, USA.

Harding, R.M. & Mills, F.J., (1988), *Aviation Medicine,* 2nd Edition, British Medical Association, London, UK.

Hawkins, F.H. (edited by Harry W. Orlady)(1993), *Human Factors in Flight,* 2nd Edition, Avebury Aviation, Aldershot, UK / Brookfield, USA.

Hurst, R. & Hurst, L., (1983), *Pilot Error - The Human Factors,* 2nd Edition, Granada Publishing, UK.

International Civil Aviation Organization (1989-), *Human Factors Bulletins.*

Joint Aviation Authority (JAA) (1991-), J*oint Aviation Requirements (JAR)*. including JAR-OPS, FAR-FCL especially (Part 3) Medical Requirements, Hoofdorp, The Netherlands.

Jensen, R.S., et al. (eds), *The International Journal of Aviation Psychology*, Laurence Erlbaum Associates, Hillsdale, NJ, USA.

Jensen, R.S., (1995), *Pilot Judgment and Crew Resource Management*, Avebury Aviation, Aldershot, UK / Brookfield, USA.

Maurino, D.E., Reason, J.T., Johnston, A.N., & Lee, R.B. (1995), *Beyond Aviation Human Factors*, Avebury Aviation, Aldershot, UK / Brookfield, USA.

Reason, J.T. & Mycielska, K., (1982), *Absent Minded? The Psychology of Mental Lapses and Everyday Errors*, Prentice-Hall Inc., New Jersey, USA.

Reason, J.T., (1990), *Human Error*, Cambridge University Press, Cambridge, UK.

Wiener, E.L. & Nagel, D.C., (1988), *Human Factors in Aviation,* Academic Press Inc., San Diego, USA.

Wiener, E.L., Kanki, B.G., and Helmreich, R.L. (eds), (1993), *Cockpit Resource Management*, Academic Press, San Diego, USA.

LIST OF ABBREVIATIONS

ASRS	Aviation Safety Reporting System
ATC	air traffic control
BMI	body mass index
CHIRP	Confidential Human Factors Incident Reporting Programme
CO	carbon monoxide
CRM	Crew/Cockpit Resource Management
CRT	Cathode ray tube
dB	decibel
DCS	decompression sickness
ECG	electrocardiogram
EEG	electroencephalogram
EMG	electromyogram
EOG	electrooculogram
FDM	Flight Deck Management
ft	foot/feet
G	unit of force related to earth's gravity
h	hour
HUD	Head-up display
ICAO	International Civil Aviation Organization
JAA	Joint Aviation Authority
LOFT	Line-Oriented Flying Training
m	metres
mm	millimetres

mmHG	millimetres of mercury
MSLT	multiple sleep latency test
N1	speed of rotation of the fan, or low pressure compressor, in a jet engine
NIHL	noise induced hearing loss
P1	captain, aircraft commander
P2	co-pilot, first officer
PAPIs	precision approach path indicators
PPB	positive pressure breathing
psi	pounds per square inch
REM	rapid eye movement
RT	radio telephony
RT	reaction time
UV	ultraviolet
V1	decision speed at take off (the speed below which the aircraft can safely be brought to a stop on the runway)
VASIs	visual approach slope indicators

INDEX